MAPS ARE TERRITORIES

SCIENCE IS AN ATLAS

A portfolio of exhibits

David Turnbull

with a contribution
by Helen Watson
with the Yolngu
Community at
Yirrkala

The University of Chicago Press

This book forms part of the HUS203/204 *Nature and human nature* course offered by the School of Humanities in Deakin University's Open Campus Program. It has been prepared by the HUS203/204 *Nature and human nature* course team.

The course includes:
Imagining nature (Study guide)
Imagining nature, Portfolio 1: *Putting nature in order*
Imagining nature, Portfolio 2: *Imagining landscapes*
Imagining nature, Portfolio 3: *Is seeing believing?*
Imagining nature, Portfolio 4: *Beasts and other illusions*
Imagining nature, Portfolio 5: *Maps are territories: science is an atlas*
Imagining nature, Portfolio 6: *Singing the land, signing the land*

These books are available from Deakin University Press, Deakin University, Geelong, Victoria, Australia 3217.

The University of Chicago Press, Chicago 60637
Deakin University, Victoria, Australia

ISBN: 0-226-81705-9 (paper)

Library of Congress Cataloguing-in-Publication Data

Turnbull, David, 1943–
   Maps are territories: science is an atlas: a portfolio of exhibits/David Turnbull; with a contribution by Helen Watson with the Yolngu community at Yirrkala. – University of Chicago Press ed. p. cm.
   Originally published: Geelong, Vic.: Deakin University, 1989.
   "Part of the HUS203/204 Nature and human nature course offered by the School of Humanities in Deakin University's Open Campus Program" – T.p. verso.
   Includes bibliographical references.
   1. Cartography – Philosophy. 2. Australia – Maps.
I. Deakin University. School of Humanities. Open Campus Program. II. Title.
GA102.3.T87 1993
912 – dc20   93 – 11239
                                                    CIP

Title page

Possibly the earliest woodcut picture of a cartographer at work (Paul Pfintzing, *Methodus geometrica*, Nürnberg, 1598)

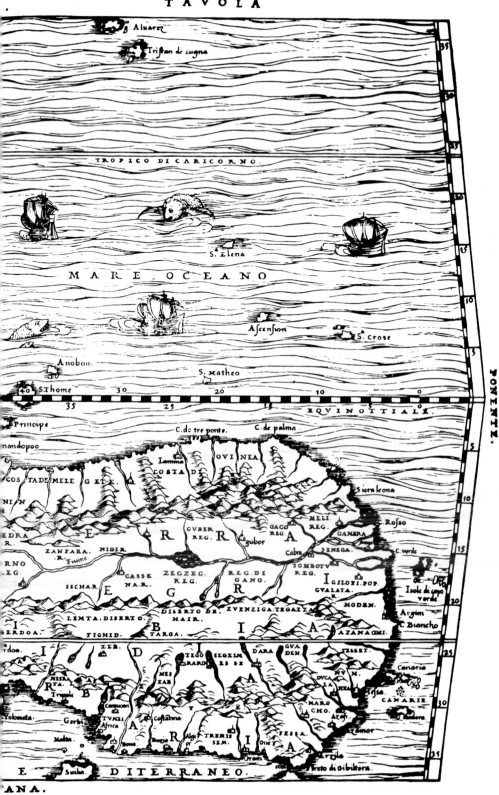

CONTENTS

◀ Africa with southward orientation. This map, published in Venice in 1556, shows the continent as it would be seen by a seafarer travelling south from Europe.

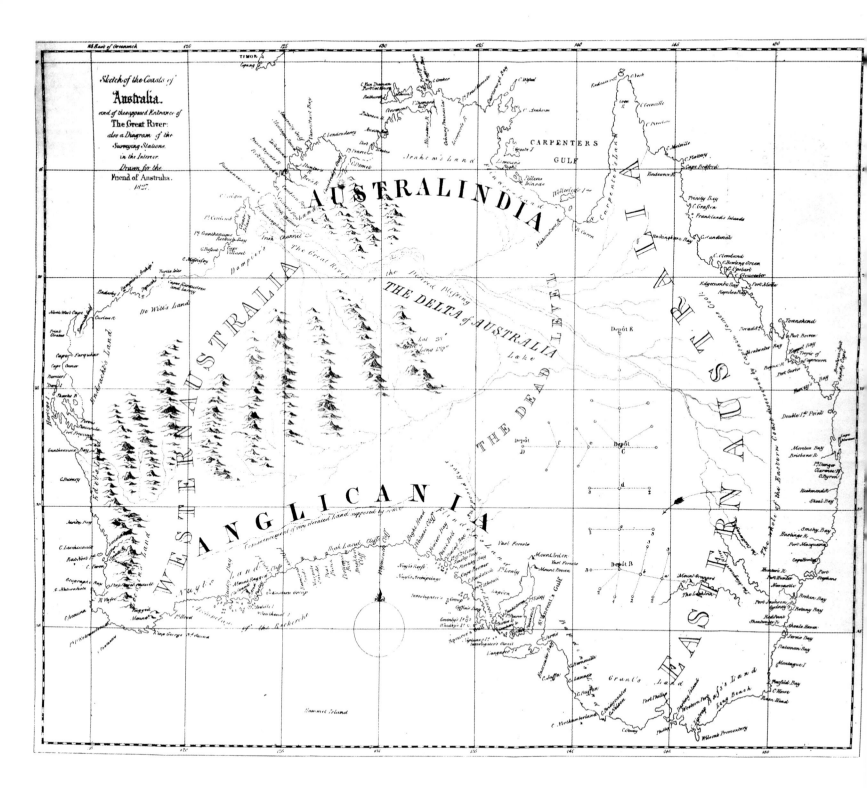

Sketch of the Coasts of
Australia.
and of the supposed Entrance of
The Great River:
also a Diagram of the
Surveying Stations.
in the Interior.
Drawn for the
Friend of Australia.
1827.

AUSTRALINDIA

CARPENTERS
GULF

WESTERN AUSTRALIA

EASTERN AUSTRALIA

THE DELTA of AUSTRALIA

THE DEAD SEA

W. ANGLICANIA

PREFACE

Two years ago, a team of three people—Wade Chambers, David Turnbull and Helen Watson—began a systematic review and critique of the cross-cultural content of the Deakin University Social Studies of Science course materials. In this book, the first of several publications resulting from that collaboration, David Turnbull analyses maps both as a metaphor for knowledge and also as a major means of knowledge representation in a wide array of cultures.

Many of the ideas presented in *Maps are territories* relate directly to other books in the *Imagining nature* series, a list of which may be found on the imprint page. Underlying all books in this series is the conviction that the great nature–culture divide is an illusion, one might almost say, a figment of the Western imagination. In attempting to define our place in the world of nature, we deal not with nature on the one hand and culture on the other but rather with many and various cultural constructions of the natural world. This is really to suggest that nature, in the experience of humanity, is not singular but manifold. Understanding nature, in this larger and more intricate sense, involves close knowledge of relevant cultural traditions.

Like the other books in this series, *Maps are territories* is conceived and structured not as a linear verbal narrative but as a progression of museum or gallery exhibits designed to exercise the skills of visualisation and visual analysis, so essential to any understanding of the basic theoretical issues of perception and cognition. A portfolio rather than a written text, each book stands alone and may be read without reference to the others. However, the full scope of the argument relating to the cultural dimensions of human perception of the natural environment will become clear only if the books are read in close conjunction.

The analysis of the interaction of European and Aboriginal knowledge systems was first articulated by Helen Watson, working as part of a group of Yolngu and non-Aboriginal Australians at Yirrkala in Arnhemland. Throughout this period of course re-evaluation, discussions were held with representatives of the Deakin University Koori Teacher Education Project, who provided funds to help keep the course development project afloat. Both the Koori Teacher Education Project and the Social Studies of Science course team supported the undertaking, with the aim of ensuring that Aboriginal knowledge receive more substantive and serious treatment in the University's curriculum as well as within the general forum of intellectual discussion.

The course team also wishes to thank several people who read and made substantive and useful comments on sections of the *Imagining nature* series: Bruno Latour, John Ziman, Nancy Williams, Barry Butcher, Chris Ryan, John Clendinnen, Kingsley Palmer, Maggie Brady and Andrew Turk.

David Wade Chambers
1989

◄ This map of Australia (1827), relatively accurate in its coastal profile, is filled with imaginary mountain ranges, rivers and deltas. Its place names, grid and topographical assumptions derive from European cultural conventions unrelated to the landscape depicted, a landscape which the Aborigines had already mapped in minute and reliable detail.

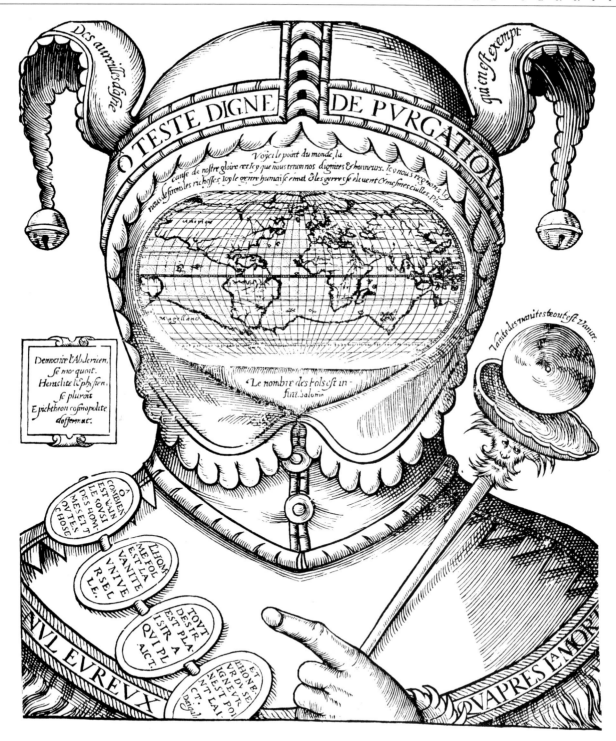

Exhibit 1
MAPS AND
THEORIES

. . . all theory may be regarded as a kind of map extended over space and time.

Michael Polanyi, *Personal knowledge: towards a post-critical philosophy*, 1958, p. 4

[In its role] as a vehicle for scientific theory, [the paradigm] functions by telling the scientist about the entities that nature does and does not contain and about the ways in which those entities behave. That information provides a map whose details are elucidated by mature scientific research. And since nature is too complex and varied to be explored at random, that map is as essential as observation and experiment to science's continuing development. Through the theories they embody, paradigms prove to be constitutive of the research activity. They are also, however, constitutive of science in other respects . . . paradigms provide scientists not only with a map but also with some of the directions essential for map-making. In learning a paradigm, the scientist acquires theory, methods, and standards together, usually in an inextricable mixture.

T. S. Kuhn, *Structure of scientific revolutions*, 2nd edn, 1970, p. 109

In these two passages Michael Polanyi and Thomas S. Kuhn equate theories with maps, and they take it for granted that the metaphor is self-explanatory. Indeed, the map metaphor is not only used to describe scientific theories, but is so pervasive that it is also commonly employed to illuminate other basic but ill-defined terms such as 'culture', 'language' and 'the mind' (see ITEM 1.1). Since metaphors play a very important role in science and in all our thinking about the world, we should be alert to the depths of constructed and construable meaning contained within them (see *Putting nature in order*, pp. 54–7). It is particularly important in this instance because there is no clear understanding amongst scientists, philosophers or cartographers as to what either a theory or a map is.

In the exhibits that follow, we shall explore the 'theory as map' metaphor by looking at maps from a range of times, places and cultures. These maps were chosen because they raise, and shed light on, a number of fundamental questions about how humans see and depict the natural world. What are maps and what is their function? What is the difference between a map and a picture? What is the relationship of the map to the landscape it represents? How do you 'read' a map?

But first let us go back to the two quotations above and ask why the map metaphor should be so persuasive and pervasive. According to the Swiss educational psychologist Jean Piaget, spatiality is fundamental to our consciousness and our understanding of experience (J. Piaget & B. Inhelder, *The child's conception of space*, 1967, p. 6ff). Most certainly, spatiality is a central element in almost all our representations of the world. Geographers Arthur H. Robinson and Barbara Bartz Petchenik explain, in the following way, this fundamental role of space in ordering our knowledge of the world:

Books in the *Imagining nature* series are referred to throughout by their titles only.

◀1.1
Court Jester with a mind as a map (1575).

1.2
Of exactitude in science

... In that Empire, the craft of Cartography attained such Perfection that the Map of a Single province covered the space of an entire City, and the Map of the Empire itself an entire Province. In the course of Time, these Extensive maps were found somehow wanting, and so the College of Cartographers evolved a Map of the Empire that was of the same Scale as the Empire and that coincided with it point for point. Less attentive to the Study of Cartography, succeeding Generations came to judge a map of such Magnitude cumbersome, and, not without Irreverence, they abandoned it to the Rigours of sun and Rain. In the western Deserts, tattered fragments of the Map are still to be found, Sheltering an occasional Beast or beggar; in the whole Nation, no other relic is left of the Discipline of Geography.

From *Travels of praiseworthy men* (1658)
by J. A. Suárez Miranda

(Jorge Luis Borges, *A universal history of infamy*, 1975, p. 131)

As we experience space, and construct representations of it, we know that it will be continuous. Everything is somewhere, and no matter what other characteristics objects do not share, they *always* share relative location, that is, spatiality; hence the desirability of equating knowledge with space, an intellectual space. This assures an organization and a basis for predictability, which are shared by absolutely everyone. This proposition appears to be so fundamental that apparently it is simply adopted a priori.

A. H. Robinson & B. B. Petchenik, *The nature of maps: essays towards understanding maps and mapping*, 1976, p. 4

Malcolm Lewis, a historical geographer, has made some interesting suggestions about the relationship between language and spatial consciousness:

Unlike the 'here and now' language of the other higher primates, human language began to bind 'events in space and time within a web of logical relations governed by grammar and metaphor'.* Wittgenstein's proposition that 'the limits of my language mean the limits of my world' remains valid.† One could go further and say that the origins of language and the growth of spatial consciousness in man are closely interrelated. The cognitive schema that underlay primitive speech must have had a strong spatial component. Not all messages were spatial in content or manifestation, but many would have been, and these would have helped to provide the structural as well as the functional foundations of language. It has been argued that these foundations helped to promote

the ability to construct with ease sequences of representations of routes and location ... Once hominids had developed names (or other symbols) for places, individuals, and actions, cognitive maps and strategies would provide a basis for production and comprehension of sequences of these symbols ... Shared network-like or hierarchical structures, when externalized by sequences of vocalizations or gestures, may thus have provided the structural foundations of language ... In this way, cognitive maps may have been a major factor in the intellectual evolution of hominids ... cognitive maps provided the structure necessary to form complex sequences of utterances. Names and plans for their combination then allowed the transmission of symbolic information not only from individual to individual, but also from generation to generation.††

M. Lewis, 'The origins of cartography', 1987, pp. 51–2.

This apparently fundamental role of space in ordering our knowledge and experience raises two different, but related, kinds of difficulties in exploring the nature of maps. Firstly it is difficult to explain the nature of maps without resorting to map-like structures in the explanation. This difficulty is a consequence of the inherent spatiality of maps, the very reason that they are so often employed as a base metaphor for language, frameworks, minds, theories, culture and knowledge. The second difficulty is that while spatiality may indeed be fundamental to all cultures, what actually counts as the 'relative location' of particular objects may not be quite so basic and may constitute one of the variables that differentiate the way cultures experience the world. That is to say, in any culture, what counts as a natural object and its spatial relations, rather than being an invariant characteristic of the world, may instead form part of that culture's world view, episteme, cognitive schema, ontology, call it what you will.

*   J. H. Crook, *The evolution of human consciousness*, Clarendon Press, Oxford, 1980, p. 148 (note 8).

†   L. Wittgenstein, *Tractatus logico-philosophicus*, tr. D. F. Pears & B. F. McGuinness, Routledge & Kegan Paul, London, 1961, para. 5.6).

†† R. Peters, 'Communication, cognitive mapping and strategy in wolves and hominids', in R. L. Hall & H. S. Sharp, *Wolf and man: evolution in parallel*, Academic Press, New York, 1978, pp. 95–107.

1.3

The Bellman's blank Ocean Chart from Lewis Carroll's *The hunting of the snark*.

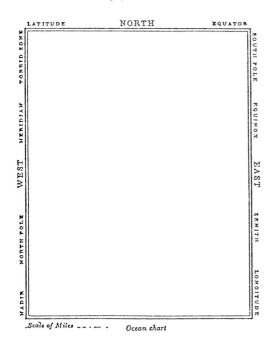

Ocean chart

The Bellman himself they all praised to the Skies—
Such a carriage, such ease and such grace!
Such solemnity too! One could see he was wise,
The moment one looked in his face!

He had brought a large map representing the sea,
Without the least vestige of land:
And the crew were much pleased when they found
it to be
A map they could all understand.

'What's the good of Mercator's North Poles and
Equators,
Tropics, Zones and Meridian lines?'
So the Bellman would cry; and the crew would
reply,
'They are merely conventional signs!

'Other maps are such shapes, with their islands and
capes!
But we've got our brave captain to thank'
(So the crew would protest) 'that he's brought *us* the
best—
A perfect and absolute blank!'

Those who are imbued with what is sometimes called 'the Western world view' think of objects as having fixed characteristics and defined boundaries (see *Putting nature in order*, pp. 48–53) and as having a position specifiable by spatial co-ordinates (see *Imagining landscapes*, p. 60). It may well be that Western ontology is in part reinforced by the centrality of maps in Western thinking and culture. Therefore, because of this possible circularity, one must be careful not to take one's own view as definitive of all maps.

There are many notoriously problematic issues, as well as some unexplored ones, bound up in such questions as 'What is the relationship between the map and the territory?' and 'When is a map not a map but a picture?'. Many of these problems are reflected in the apparent cogency of Korzybski's dictum 'The map is not the territory' (*Science and sanity*, 1941, p. 58). After all, if the map were identical with the territory it would literally *be* the territory. It would have a scale of an inch to the inch and, apart from anything else, it would be unworkable as a map since you would have to be standing on it or in it. Lewis Carroll described such a map in *Sylvie and Bruno concluded*. In this fantasy, a Professor explains how his country's cartographers experimented with ever larger maps until they finally made one with a scale of a mile to a mile. 'It has never been spread out, yet', he says. 'The farmers objected: they said it would cover the whole country, and shut out the sunlight! So now we use the country itself, as its own map, and I assure you it does nearly as well.'

Two general characteristics of maps emerge from such seemingly whimsical examples as Jorge Luis Borges's cartographic empire (see ITEM 1.2) and the Bellman's blank chart (see ITEM 1.3). Firstly, maps are *selective*: they do not, and cannot, display all there is to know about any given piece of the environment. Secondly, if they are to be maps at all they must directly represent at least *some* aspects of the landscape.

We may divide the types of representation in maps into two different types: iconic representation (which attempts to directly portray certain visual aspects of the piece of territory in question) and symbolic representation (which utilises purely conventional signs and symbols, like letters, numbers or graphic devices). For example, look at ITEM 6.1 and try to distinguish those elements of the map which are representational (iconic) from those which are entirely reliant on arbitrary convention (symbolic).

J. B. Harley and David Woodward have recently proposed an all-embracing definition of maps: 'Maps are graphic representations that facilitate a spatial understanding of things, concepts, conditions, processes, or events in the human world' (J. B. Harley & D. Woodward (eds), *The history of cartography*, vol. 1, 1987, p. xvi). For our purposes we can take a working definition of a map as a graphic representation of the milieu, containing both pictorial (or iconic) and non-pictorial elements. Such representations may include anything from a few simple lines to highly complex and detailed diagrams.

Exhibit 2
THE CONVENTIONAL NATURE OF MAPS

We said in Exhibit 1 that a map is always selective. In other words, the mapmaker determines what *is*, and equally importantly, what *is not* included in the representation. This is the first important sense in which maps are *conventional*. What is on the map is determined not simply by what is in the environment but also by the human agent that produced it. Furthermore, we saw in Exhibit 1 that maps employ non-iconic signs and symbols. These are as arbitrary as the letters of the alphabet and are therefore largely conventional. Of course, many elements of maps are at least partly iconic, portraying certain visual features of the landscape represented, but even these images partake to a significant degree of the conventions of the artist (see *Imagining nature*, pp. 35–8, and *Beasts and other illusions*, pp. 24–39).

The historian of geology Martin Rudwick has discussed the inherent conventionality of maps in the context of his argument that geology could not become a fully developed science before the development of visual diagrams:

> . . . a geological map . . . is a document presented in a visual language; and like any ordinary verbal language this embodies a complex set of tacit rules and conventions that have to be learned by practice. . . [Therefore there also has to be] a social community which tacitly accepts these rules and shares an understanding of these conventions.
>
> Martin Rudwick, 'The emergence of a visual language for geological science, 1760–1840', 1976, p. 151

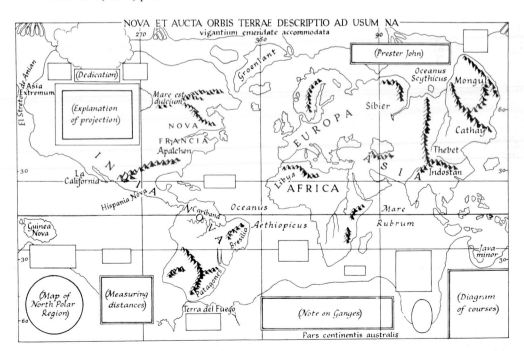

◀ **2.1**
Mercator's projection maintains compass directions between places as straight lines. This is achieved by making the distance between lines of latitude greater towards the poles and thus introducing considerable distortion of area and shape. The boxed areas indicate that Mercator had an understanding of the epistemological status of a map that differs from our own. Some of the boxes contain more information about how to use the map than we now deem necessary: how to measure distances, for example; and the box in the north-east corner of the map is about Prester John, the mythical king of Africa.

2.2
The Peters projection maintains
north-south and east-west directions
and preserves the relative size of
countries at the expense of shape.

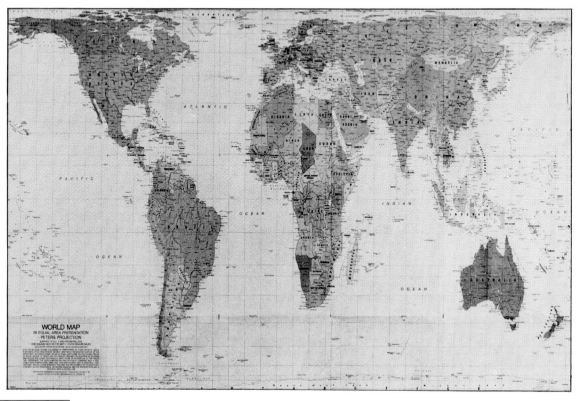

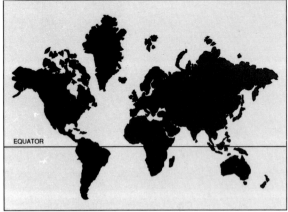

2.3
The distorted area effect of Mercator's projection.
The area of Greenland is 2 176 000 square
kilometres; the area of Australia is 7 690 000 square
kilometres.

Another obvious way in which maps are conventional is in their use of 'projection'. No curved surface like that of the Earth can be projected in two dimensions without some distortion. Over the years many different modes of projection have been developed: some are better for conveying such elements as shape or size; some, for compass direction or relative position; some are more distorted toward the poles; some, towards the equator. No one projection is the best or the most accurate. A particular projection is selected by the mapmaker on the basis of functional and perhaps aesthetic criteria, or because of a specification or convention.

The projection developed by Gerhard Mercator, a Flemish cartographer, in 1569 became the most commonly used projection because it portrayed compass directions as straight lines. However, this was achieved at the expense of distortion of relative size, especially towards the poles. Mercator's Greenland appears much larger than Australia, which is in reality more than three times the size of that North Atlantic island. But other more subtle effects result from Mercator's view of the world. If you compare the Mercator projection (ITEM 2.1) with the Peters projection (ITEM 2.2), a map which endeavours to preserve relative size, what differences do you discover which might have cultural or political significance? You may wish to ask yourself what interests are served in a Mercator projection. Is it a coincidence that a map which preserves compass direction (a boon for ocean navigation) shows Britain and Europe (the major sea-going and colonising powers of the past 400 years) as relatively large with respect to most of the colonised nations? (See ITEM 2.3.) What if we turn the Peters projection upside down and centre it on the Pacific? (See ITEM 2.4.) A profoundly altered view of the world is obtained.

2.4
The Peters projection inverted and recentred.

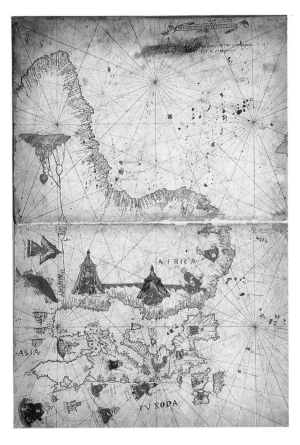

2.5
Africa with south at the top, reflecting the
perspective of the explorers of the day (Vesconte
Maggiolo, 1512).

Of course, orientation is an arbitrary convention (see ITEM 2.5 and Contents page).
Indeed, the very word *orient*-ation comes from 'East' being the direction of the rising sun
and hence it was once common practice to put it at the top of the map. North, whilst
being one end of the Earth's axis of rotation, is not a privileged direction in space,
which after all has no 'up' or 'down'. That North is traditionally 'up' on maps is the
result of a historical process, closely connected with the global rise and economic
dominance of northern Europe.

Another early example of map orientation reveals several interesting points (see ITEM 2.6).
This map, produced less than 100 years after the Spanish conquest of the Inca, was
drawn by the Quechua writer Hawk Puma (Guaman Poma) to illustrate his moving
account of Spanish misrule of Peru. Because his manuscript was part of a petition to the
Spanish monarch, Hawk Puma employed a number of *European* conventions: he called it
Mappamundi; he put 'North' at the top; he added pictorial elements such as the sun,
ships, mountains, buildings in the towns and cartographically familiar sea creatures. The
map shows little connection with the sophisticated relief maps which the 17th-century
historian Garcilaso de la Vega attributes to the Inca, and it thus appears to avoid Incan
cartographic conventions in favour of the European.

To modern eyes, the Hawk Puma map may at first seem fairly 'primitive', bearing, as
commentators have suggested, only limited relation to the actual landscape depicted. For
example, neighbouring countries seem to be wrongly placed, with the Pacific to the
south instead of the west. But what happens if we rotate the map? Suddenly all the
geographical relations fall into place. Even the rivers, the Maranon, Amazon and
Pilcomayo begin to flow in the right directions again.

The significance of this rotation is much more interesting than a simple factual error.
We now see that Hawk Puma's map follows long-standing *Incan* conventions after all.
The explanation is historical. When the Inca expanded into Chile, a mountain pass had
taken them by way of the great 'Eastern' society of the Collasuyu. Thus, Chile, to the
south, came to be considered by the Inca as an *eastern* extension of their empire.
Colombia, to the north, for similar reasons was considered a *western* extension.
Furthermore, the map centres on Cuzco, the Inca capital, rather than on Lima, the
Spanish capital; and its extent approximates the geographical limits of the Incan empire
in South America, not the limits of the Spanish Indies, which the map purports to
depict.

Thus, the story of the Hawk Puma map teaches us that conventions often follow
cultural, political and even ideological interests, but that if conventions are to function
properly they must be so well accepted as to be almost invisible. The map, if it is to
have authority in Western society, must have the appearance of 'artless-ness'; that is, it
must appear simply to exhibit the landscape, rather than to describe it with artifice or in
accordance with the perceived interests of the mapmaker. For a map to be useful, it

2.6
Hawk Puma's map (1613)

Sketch map of the west coast of South America showing the extent of the Inca empire before the arrival of the Spanish. Also shown are the approximate relations to Cuzco of the four neighbouring cultural groups which are identified on the four sides of Hawk Puma's map. Note that when rotated counterclockwise until the Condesuyu are south of Cuzco, this map approximates Hawk Puma's orientation.

must of course offer information about the real world, but if this 'real world information' is to be credible, it must be transmitted in a code that by Western standards appears neutral, objective and impersonal, unadorned by stylistic device and unmediated by the arbitrary interests of individuals or social groups. (We frequently use maps that are highly stylised though: for example, guide maps to places and institutions.) Hawk Puma's map was almost certainly never seen by the King of Spain, and his petition never read. It may be that this failure resulted at least in part from the visibility of the Incan conventions, which would have been read by the Spanish eye as inaccuracies, thus calling the map's credibility into doubt.

In this regard, maps prove once again an apt metaphor for scientific discourse. Scientific representations of the phenomenal world are, like maps, laden with conventions, which are kept as transparent, as inconspicuous, as possible. This has, in the past generation, been brought home to us through the work of many commentators. Scientists, such as Peter Medawar, have described how scientific papers systematically conceal actual scientific practice ('Is the scientific paper a fraud?', 1963). Philosophers and sociologists of science have pointed out that though we may grant that there *is* an external material world, we can gain no direct, unmediated experience of it. This obviously makes for a considerable difficulty. If there is a material world, and we have no direct experience of it, what is the relationship of our representations of the world to that world? Philosophers and sociologists who have taken a constructivist approach hold that rather than accept a split between the world and our experience of it, we should consider our representations of the world as active constructions. On this view, our experience of the world and our representations of it are mutually interdependent, so there is a sense in which the two are inseparable. Or, to put it in its most contentious form, 'the map is the territory'.

An analytic device which may guide us in looking at these complex and contentious issues is Ludwig Wittgenstein's 'forms of life', by which he means that all language, communication and shared experience has to be based in doing, in practical action (L. Wittgenstein, *Philosophical investigations*, 1953, Book 2, xi, p. 226). Steven Shapin and Simon Schaffer define a 'form of life' as the existing scheme of things, the invisible, conventional and self-evident 'patterns of doing things and of organizing men to practical ends' (S. Shapin & S. Schaffer, *Leviathan and the air-pump*, 1985, p. 15), and it is in this sense that we will be using the term. A form of life can be taken as a set of conventional linguistic practices and social structures that are 'given', without which there can be no talk, knowledge or social relations. These 'givens' structure what it is possible to ask and what it is possible to answer. They lay down the criteria for what is to count as knowledge. From this constructivist perspective, knowledge can be seen as a practical, social and linguistic accomplishment, a consequence of the bringing of the material world into the social world by linguistic and practical action.

Look at ITEM 2.7, an intergalactic map, devised by the astronomer Carl Sagan, showing humans and their universe. Is the map as culture free as it was intended to be? Can we be sure that any intelligent being from another galaxy could read it? Can one find the 'forms of life' that reside in this representation? Similarly, consider ITEM 2.8, the map of the London Underground. This map varies scale and position in order to display the configuration and interconnections of the rail network. It is designed to be useful to all users of the Underground, including members of other than European cultures. Does it succeed in this? What 'forms of life' underlie this map? What does it sacrifice? Is there a sense in which it is art?

It is apparent from the above description of forms of life that they are closely related to what Kuhn described earlier as the map-like character of theories. In the words of a philosopher of science who brings maps and forms of life together:

> To talk, in the philosophy of science, of theoretical physics falsifying by abstraction, and to ask for the facts and nothing but the facts, is to demand the impossible, like asking for a map to be drawn to no particular projection and having no particular

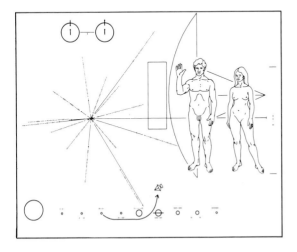

2.7
This metal plaque was designed by the American astronomer Carl Sagan and placed aboard the Pioneer 10 spacecraft currently on its way out of our solar system.

2.8
The London Underground (Henry Beck, 1931).

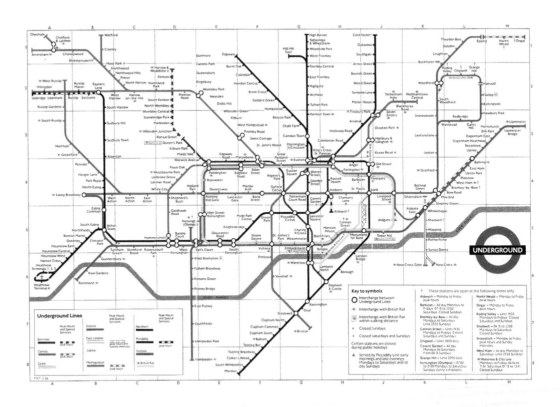

scale . . . If we are to say anything, we must be prepared to abide by the rules and conventions that govern the terms in which we speak: to adopt these is no submission nor are they shackles. Only if we are so prepared can we hope to say anything true— or anything untrue.

Stephen Toulmin, *The philosophy of science*, 1953, p. 129).

Or in Stuart Hall's words 'you cannot learn, through common sense, *how things are*: you can only discover *where they fit* into the existing scheme of things' (S. Hall, 1977, in D. Hebdige, *Subculture: the meaning of style*, 1979, p. 11).

Maps and forms of life are obviously closely interwoven notions. So let us try and make their relationship a little clearer by looking at the question of how we know that something is a map and not a picture.

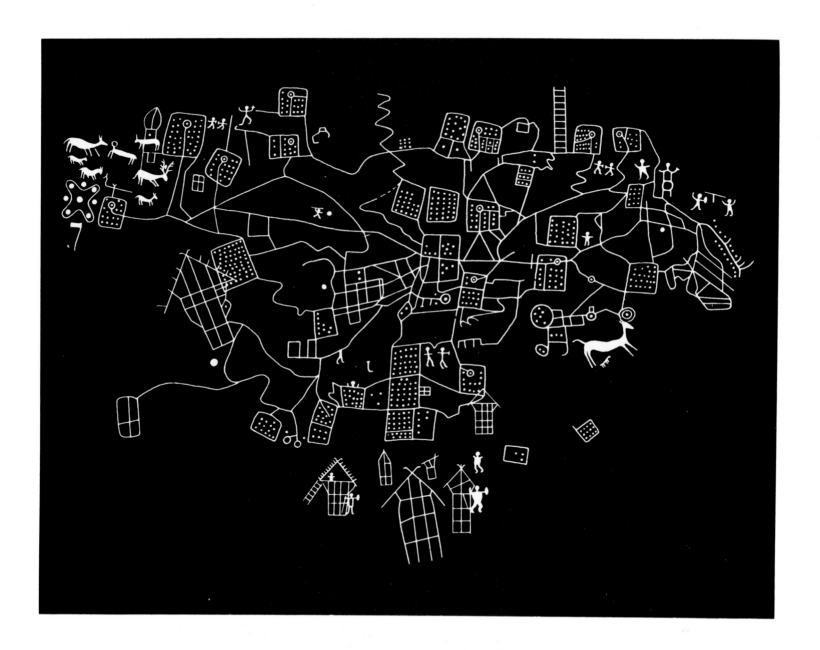

3.1
Bedolina petroglyph, Valcamonica.

Exhibit 3
MAPS AND
PICTURES

Before we attempt to discuss the difference between maps and pictures, we need to look again at the question 'What is a map?'. It is claimed that the Bedolina and Giadighe petroglyphs at Valcamonica (2500 BC) (ITEMS 3.1 and 3.2), and the wall painting at Çatal Hüyük (6200 BC) (ITEM 3.3) are amongst the oldest examples of maps. Exactly how old they are is really not important since they are old enough that we have no direct knowledge of the culture in which they were made. What is it that makes it reasonable to claim they are maps? What do we mean when we say they look like maps?

They appear to portray a particular landscape. They have a partly plan-like character, that is, they seem to have a bird's-eye viewpoint. They appear to be only partly iconic, having some symbolic elements with a degree of regularity. We can read the petroglyphs as showing paths, fields, houses and people. Beyond this it is difficult to speculate, since we have no clue as to the purposes of those who drew them. The question of purpose seems crucial, because we would be less willing to call them maps if they clearly had a pictorial, religious, ritual, symbolic or magical function. Yet these different functions need not be incompatible.

3.2

Giadighe petroglyph, Valcamonica.

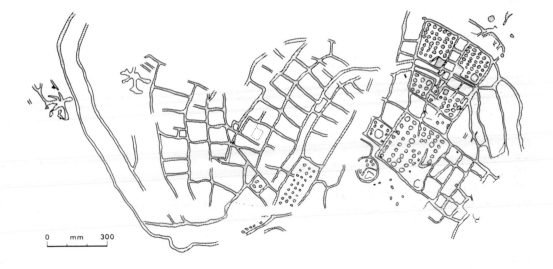

0   mm   300

3.3

Wall painting at Çatal Hüyük.

3.4
Nippur plan (1500 BC)

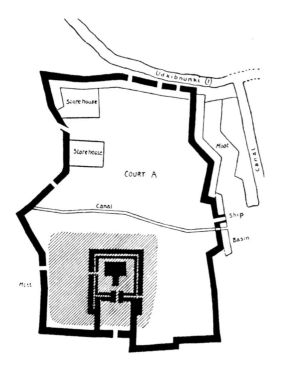

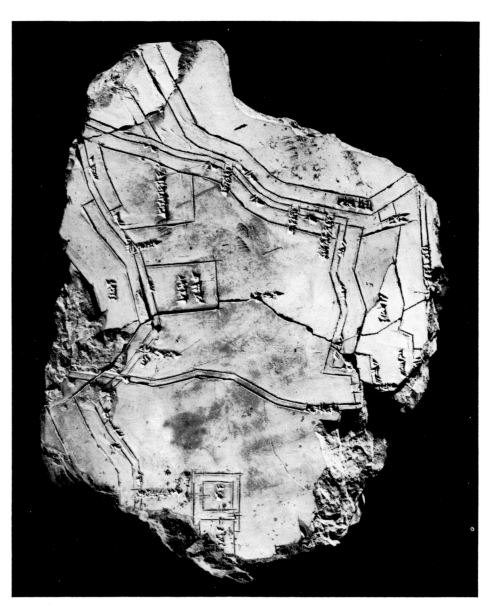

In the case of the two clay tablets from Nippur (1500 BC) and Nuzi (2300 BC) (ITEMS **3.4** and **3.5**), though they are of great antiquity, we assume that the cultural continuity and similarities with our own notions of maps are sufficient to identify them as such. They appear to have the clear purpose of representing an identifiable piece of landscape. What features on the tablets would lead to this conclusion?

'What a picture is' is probably one of those deceptively simple questions that philosophy can never answer, but perhaps we can settle on a couple of points. Many pictures are presumably representations of a particular subject or part of the landscape from a particular point of view. The point of view is taken as having at least some significance and may indeed be the dominant aspect of the picture. Whereas maps, though they have a point of view in the sense that they are representations of parts of the landscape, deny or suppress that point of view. This is one of the conventions which we described in Exhibit 2 as 'transparent'. Maps have been thought to be objective in that they are independent of the view of a particular observer. This reveals another of the reasons that theories are held to be analogous to maps:

> Theoretical understanding is supposed to disengage us. Theoretical understanding is nonperspectival and therefore treats all locations in space or time as theoretically equivalent (it allows no epistemological privilege to any spatiotemporal framework).
>
> Joseph Rouse, *Knowledge and power: toward a political philosophy of science*, 1987, p. 70

However, as we shall see, this notion of maps as non-perspectival representations will not do. It is not just that maps do have a perspective, or that the perspective is taken for granted, it is rather that the disengagement hides the privileging of a particular conceptual scheme. Maps, in this sense, are pictures. They are pictures with different and additional functions and purposes to those of perspectival representation. Pictures may sometimes be entirely subjective, but maps, to be capable of transmitting information, have to be intersubjective.

3.5
Clay tablet from Nuzi (2300 BC)

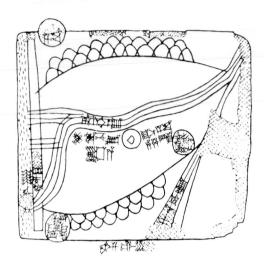

Look now at three relatively contemporary examples (ITEMS **3.6**, **3.7** and **3.8**), which are not easily distinguished from pictures and clearly have aesthetic qualities as well as practical ones. What elements in ITEMS **3.6** and **3.7** would you say make them maps, and what elements give them the character of pictures? Is the Leonardo da Vinci example (ITEM **3.8**) neither a map nor a picture? Why not? Is it a diagram? What is it about the Enniskillen map (ITEM **3.9**) that makes it a rather doubtful example of a map by Western standards?

3.6
Verona, Lake Garda and the Adige valley (mid-15th century).

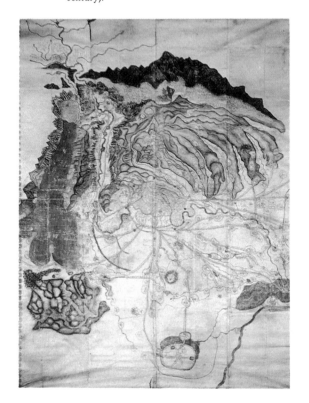

3.7
Kingsai province south-east China (18th century).

3.8
Plan by Leonardo da Vinci to regulate the river
Arno (1502).

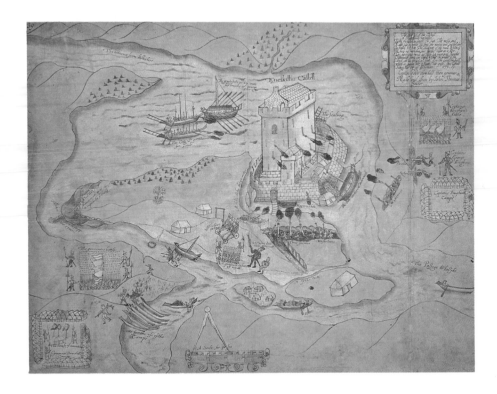

3.9
Siege of Enniskillen (1594).

1.1

Chippewa Indian land claim presented to the US Congress in 1849.

This is the leading inscription, and symbolizes the petition to the President.
No. 1. It commences with the totem of the chief, called Oshcabawis, who
headed the party, who is seen to be of the *Ad-ji-jauk*, or Crane clan. To the eye
of the bird standing for this chief, the eyes of each of the other totemic animals
are directed as denoted by lines, to symbolize *union of views*. The heart of each
animal is also connected by lines with the heart of the Crane chief, to denote
unity of feeling and purpose. If these symbols are successful, they denote that the
whole forty-four persons both *see* and *feel* alike—THAT THEY ARE ONE.
No. 2, is a warrior, called Wai-mit-tig-oazh, of the totem of the Marten. The
name signifies literally, He of the wooden Vessel, which is the common
designation of a Frenchman, and is supposed to have reference to the first
appearance of a ship in the waters of the St. Lawrence.
No. 3. O-ge-ma-gee-zhig, is also a warrior of the Marten clan. The name
means literally, Sky-Chief.
No. 4, represents a third warrior of the Marten clan. The name of Muk-o-mis-
ad-ains, is a species of small land tortoise.

No. 6, *Penai-see*, or the Little Bird of the totem of the *Ne-ban-a-baig*, or Man-
fish. This clan represents a myth of the Chippewas, who believe in the existence
of a class of animals in the Upper Lakes, called *Ne-ban-a-baig*, partaking of the
double natures of a man and a fish—a notion which, except as to the sex, has its
analogies in the superstitions of the nations of western Europe, respecting a
mermaid.
No. 7. *Na-wa-je-wun*, or the Strong Stream, is a warrrior of the O-was-se-wug,
or Catfish totem.
Beside the union of eye to eye, and heart to heart, above depicted, Osh-ca-ba-
wis, as represented by his totem of the Crane, has a line drawn from his eye
forward, to denote the course of his journey, and another line drawn backward
to the series of small rice lakes, No. 8, the grant of which constitutes the object
of the journey. The long parallel lines, No. 10, represent Lake Superior, and the
small parallel lines, No. 9, a path leading from some central point on its
southern shores to the villages and interior lakes, No. 8, at which place the
Indians propose, if this plan be sanctioned, to commence cultivation and the arts
of civilized life. The entire object is thus symbolized in a manner which is very
clear to the tribes, and to all who have studied the simple elements of this mode
of communicating ideas.
(H. R. Schoolcraft, *Historical and statistical information respecting the history,
condition and prospects of the Indian tribes of the United States*, part 1, 1851,

Exhibit 4
BRINGING THE WORLD BACK HOME

The maps in this exhibit are from societies that were once called 'primitive', and as such they are supposed to stand in marked contrast with technically accurate maps of contemporary Western society. Malcolm Lewis, who has written extensively on American Indian maps, points out that they 'differed from post-Renaissance European maps in two fundamental respects: geometrical structure and the selection and ordering of information content'. European maps have a projective geometry based on a co-ordinate system. Indian maps are topologically structured 'conserving connectivity between the parts but distorting distance, angles and, hence, shape'. European maps have standardised representation, but Indian maps served specific functions in particular contexts (M. Lewis, 'Indian delimitations of primary biogeographic regions', 1987, p. 94).

It is often argued that maps are scientific and that what makes them so is that they embody, as does science, statements that are true, independent of the context in which they are made (for example, $E = MC^2$). Such statements are called non-indexical. Indexical statements are those that are dependent for their truth on their context. For example, the Chippewa Indian land claim presented to Congress in 1849 (ITEM 4.1) is recogniseably a map, but the information it conveys can only be understood within the cultural specifics of the circumstances that it portrays and cannot be generalised beyond that context. That so-called 'primitive' maps serve specific functions in particular contexts clearly makes them indexical, though ITEM 4.2 rather ironically shows that context boundaries may be transgressed quite readily on occasions. The temptation is to assume that modern projective maps are non-indexical. This would mean both that the

I really wonder about that.

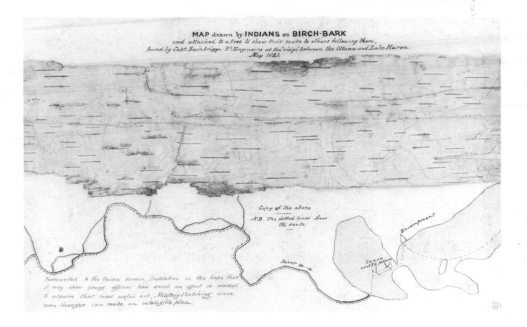

4.2
'Map drawn by Indians on Birch-Bark' (1841). Note the accompanying sketch and note. The note reads: 'Forwarded to the United Service Institution in the hope that it may shew young officers how small an effort is needed to acquire that most useful art, Military Sketching, since even Savages can make an intelligible plan.'

4.3
Stick-charts from the Marshall Islands. The shells represent islands, and the sticks represent currents and lines of swell. Such charts were used for instruction in learning to navigate rather than as navigational aids.

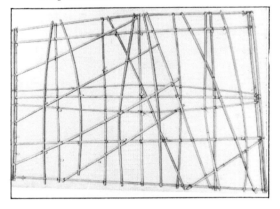

position of objects on such a map could be ascertained without reference to a point of view, and that statements about their position could be read directly off the map, without any exposure to the forms of life in which they are embedded. That is to claim that they could be understood independently of their context of use, the world view, cognitive schema or the culture of the mapmaker. In this exhibit, we suggest that this distinction is overdrawn, and that all maps are in some measure indexical, because no map, representation or theory can be independent of a form of life.

In order to understand why it is that all maps are indexical, let us first consider some examples of early maps: the Marshall Island stick-charts (ITEM 4.3) and the Inuit coastal chart (ITEM 4.4). Without a full understanding of the forms of life in which they are embedded, we cannot read them, though for their makers they provide useful information. Red Sky's migration chart (ITEM 4.5), though distorted and indexical, is clearly readable once we compare it with a Western map (ITEM 4.6). Non Chi Ning Ga's Missouri map (ITEM 4.7) is an American Indian map which differs from a modern map of the same area (ITEM 4.8) only in the details. These examples show that so-called 'primitive' maps are in fact comparable with modern Western maps in many respects.

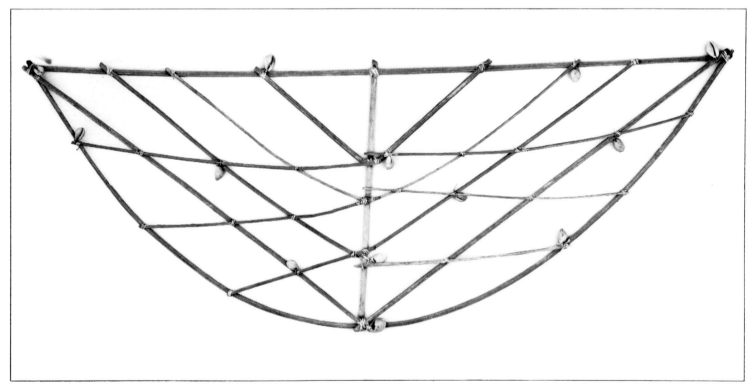

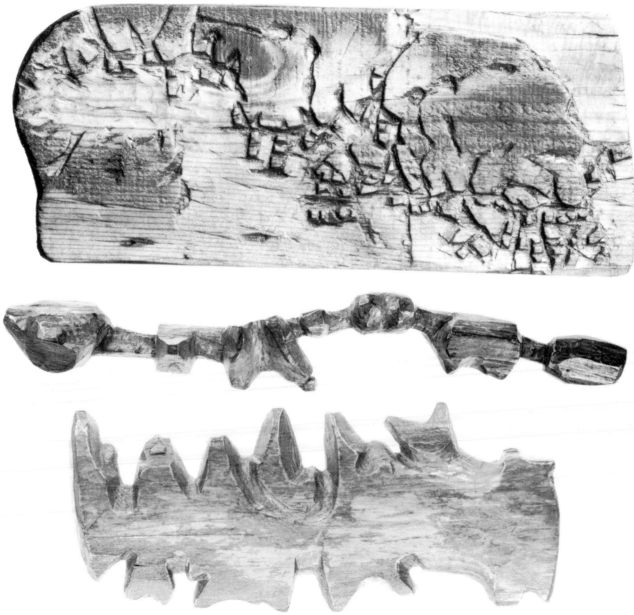

4.4
Carved wooden coastal charts carried in their kayaks
by Greenland Inuit (Eskimo). The middle two form
a single map: the shorter piece represents a stretch
of coast and the larger, islands offshore. Both are
read continously along each side.

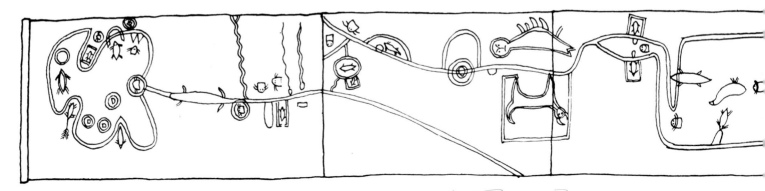

4.5
Red Sky's migration chart drawn on a birchbark
scroll 2.6 m long. It portrays the migration of the
southern Ojibway Indians in mythical times.

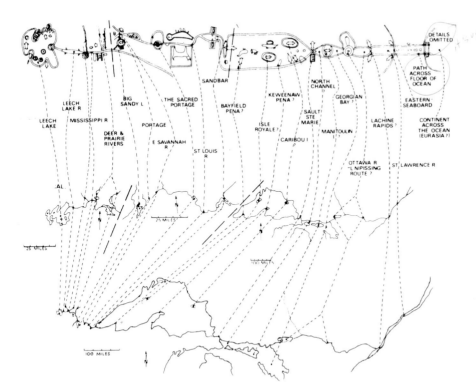

4.6
Contemporary Western geographical interpretation
of Red Sky's chart.

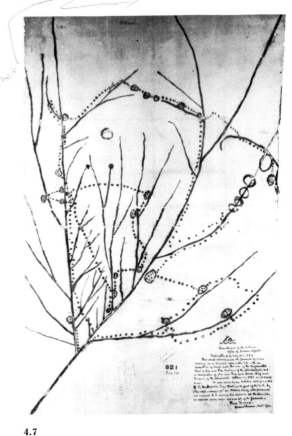

4.7
Manuscript map of the upper Mississippi and lower
Missouri presented by Non Chi Ning Ga, an Iowa
Indian chief, as part of a land claim in Washington,
1837.

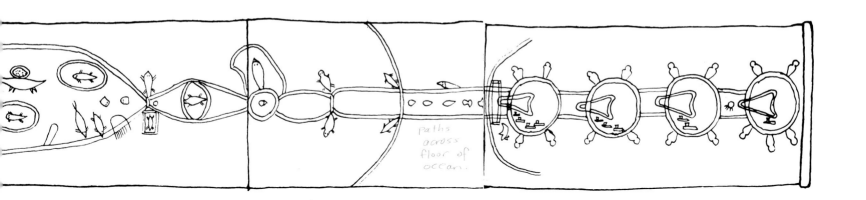

Paths
across
floor of
ocean.

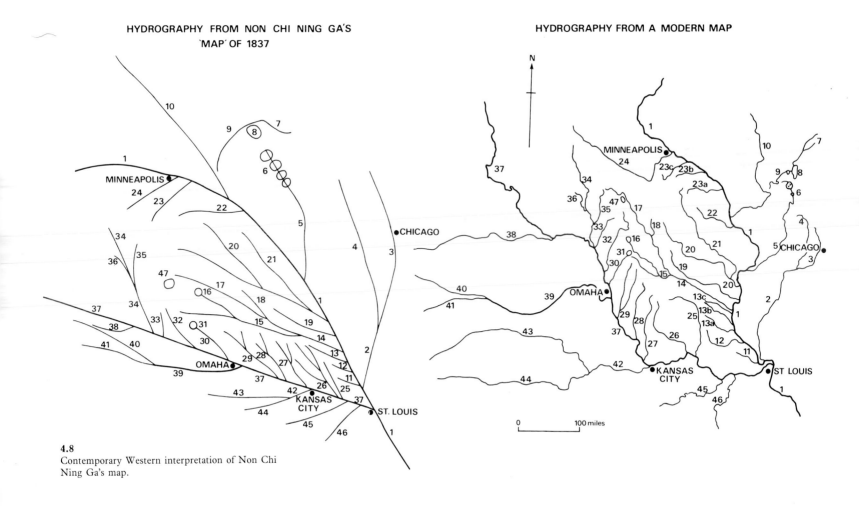

HYDROGRAPHY FROM NON CHI NING GA'S
`MAP` OF 1837

HYDROGRAPHY FROM A MODERN MAP

4.8
Contemporary Western interpretation of Non Chi
Ning Ga's map.

4.9
Pencil drawing of the Belcher Islands by the Inuit
Wetalltok.

The ability of indigenous peoples to draw accurate maps is also shown in the example of Wetalltok's map of the Belcher Islands (ITEM **4.9**). White explorers of Hudson's Bay were unsure of the existence of the Belcher Islands and were somewhat sceptical of the Inuit claims about them (see ITEM **4.10**). They were put on Western maps merely as a matter of guesswork until Wetalltok drew a map of them in 1895. Eventually Flaherty's expedition in 1912–16 established the accuracy of Wetalltok's map of a very complex bit of topography (see ITEM **4.11**).

It would seem, then, that many apparently 'primitive' maps are just as capable of conveying useful information as are Western maps. So let us now return to the question of whether Western maps are non-indexical by examining the way they structure information using a projective geometry based on a co-ordinate system. The introduction of perspective geometry in Renaissance Europe had a revolutionary impact:

> Following the discovery of perspective geometry, the position of man in the cosmos altered. The new technique permitted the world to be measured through proportional comparison. With the aid of the new geometry the relative sizes of different objects could be assessed *at a distance* for the first time. Distant objects could be reproduced with fidelity, or created to exact specifications in any position in space and then manipulated mathematically. The implications were tremendous. Aristotelian thought had endowed all objects with 'essence', an indivisible, incomparable uniqueness. The position of these objects was, therefore, not to be compared with that of other objects, but only with God, who stood at the centre of the universe. Now, at a stroke, the

4.10
Map of Hudson Bay and Belcher Islands before
Flaherty's expedition (1912–16).

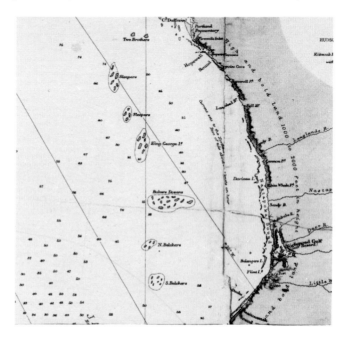

4.11
Map of Belcher Islands after Flaherty's expedition,
confirming the accuracy of Wetalltok's map.

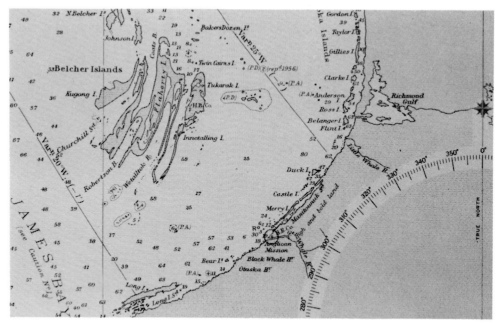

special relationship between God and every separate object was removed, to be
replaced by direct human control over objects existing in the same, measurable space.
   This control over distance included objects in the sky, where the planets were
supposed to roll, intangible and eternal, on their Aristotelian crystal spheres. Now
they too might be measured, or even controlled at a distance. Man, with his new
geometrical tool, was the measure of all things. The world was now available to
standardisation. Everything could be related to the same scale and described in terms
of mathematical function instead of merely its philosophical quality. Its activity could
also be measured by a common standard, and perhaps be seen to conform to rules
other than those of its positional relationship with the rest of nature. There might
even be common, standard, measurable laws that governed nature.

James Burke, *The day the universe changed*, 1985, pp. 76–7

It was not until the early 1400s that Ptolemy's *Geographia* arrived in Europe, the same
period in which Brunelleschi developed perspective geometry and its application in
architecture. The *Geographia* mapped the entire world and presented all the known
information in a standardised and consistent way with grid lines of latitude and
longitude (see ITEM 7.1). This metrication meant that all points were commensurable: that
is, distances and directions could be established between one place and any other.
Further, unknown places could be given co-ordinates. It was the synthesis of perspective
geometry and Ptolemy's work that enabled the imposition of a grid on the known world.
Once that grid was imposed, the mathematican Toscanelli was able to argue plausibly
that sailing westwards across the Atlantic was a shorter voyage to the Spice Islands than
the traditional route around the Cape of Good Hope and on to the East. Thus

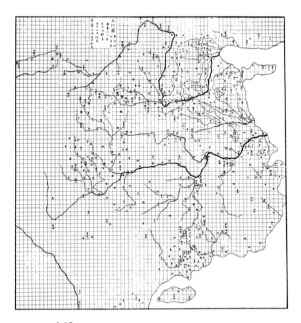

4.12
Chinese grid map drawn on stone, known as 'Map of the tracks of Yu the Great' (1137).

Columbus 'discovered' America even though in 1492 he was convinced that Cuba and Japan were one and the same.

There is of course nothing in reality that corresponds to such a grid; it is a human construct, and hence arbitrary, conventional and culturally variable. Ptolemy (AD 90–168) located his grid by designating the Fortunate Isles (Canary Islands) as the prime meridian, because they were the western extremity of the known world. Spanish and Portuguese cartographers used the Tordesillas line (see ITEM 10.2). The use of grids originated in China, probably with the work of Chang Heng in the 1st century AD. Although none of his mapwork survives, his biographer, Tshai Yung, wrote that he 'cast a network about heaven and earth and reckoned on the basis of it' (R. Temple, *The genius of China*, 1989, p. 30). Subsequently the grid was in continuous use in China, and one of the two early maps inscribed in stone at Sian in 1137 is covered by a well-defined grid (ITEM 4.12). Phei Hsiu, the 3rd-century Chinese cartographer, laid down the use of the grid as one of his six principles of scientific cartography, claiming that 'When the principle of the rectangular grid is properly applied, then the straight and the curved, the near and the far, can conceal nothing of their form from us' (in P. D. A. Harvey, *The history of topographical maps*, 1980, pp. 133–4). The Romans used a grid system called 'centuration' by which they came close to turning 'all Europe into one vast sheet of graph paper' (S. Y. Edgerton, 'From mental matrix to *Mappamundi* to Christian Empire', 1987, p. 22). The British use the National Grid system and in Australia the Australian Map Grid is used.

Even the system of lines of latitude and longitude are conventional: 'At an international conference held in Washington in 1884 it was agreed by many countries that subsequently 0° of longitude would be assumed to pass through Greenwich [England]. This meridian is now widely, though not universally, used for mapmaking' (A. G. Hodgkiss, *Understanding maps*, 1981, p. 30). For a grid system to work it has to be *literally* conventional. Grid systems require real conventions, negotiations and agreements. In order to bring the distant and the large to your table top you need perspective geometry, reproducible and combinable representations, a grid and the agreement of your fellows. The power of maps lies not merely in their accuracy or their correspondence with reality. It lies in their having incorporated a set of conventions that make them combinable in one central place, enabling the accumulation of both power and knowledge at that centre. The significance of Ptolemy's *Geographia* was not just its use of a grid: it was also an atlas which enabled the co-ordination of maps of individual lands into one map of the world. Similarly, the map of China (ITEM 4.12) was constructed as a pathwork of local maps drawn from itineraries.

In Exhibit 5 we shall be considering Aboriginal Australian bark paintings as maps. These have the appearance of being incapable of being combined in the European or Chinese way. Their maps appear to have no grid, no standardised mode of representation. Nonetheless it is possible for Aboriginal people to know about, and to travel across, unknown, even distant, territory. Their knowledge is in fact combinable because it is in the form of narratives of journeys across the landscape. Aborigines inculcate and invoke conventions just as we do, through conferences and agreement. They call them business meetings; anthropologists call them ceremonies and rituals. Songlines (which are accounts of journeys made by Ancestral Beings in the Dreamtime) connect myths right across the country. One individual will only 'know' or have responsibility for one section of the songline, but through exchange and negotiation, the travels of the Ancestors can be connected together to form a network of dreaming tracks. These may be constituted as bark paintings or song cycles.

The strength of the distinction between indexical maps and non-indexical maps will seem even less cogent when we come to consider the use of maps as instruments for navigation in Exhibit 9. It will then be seen that in order to find our way about we need at least a mental map, or a cognitive schema, and an indexical image of the landscape, and that we can never navigate with non-indexical statements alone.

That maps consisting entirely of non-indexical statements cannot be used for practical purposes involving direct interaction with the material world such as navigation constitutes another similarity with scientific theories. Scientific theories consist of non-indexical universal statements about reality, and as such cannot be applied directly to a particular circumstance, or be confirmed or falsified by particular items of evidence. Scientific theories always need additional assumptions and qualifications, or specification of conditions in order to apply in practice. Strictly speaking, scientific theories could be said to be, on the one hand, useless, or, on the other hand, false, in the sense that they can never apply without modification to a particular circumstance. If, for example, you wish to calculate the orbit of a planet around the sun, Newton's laws of motion and gravitation, or even Einstein's laws, are not sufficient. You have to assume that there are no other forces at work, that there are no unobserved bodies in our solar system, that space is empty, that planets are effectively point masses, that there are no effects of the system acting on itself and so on. Then you have to live with the fact that in the case of Mercury, for example, the predicted orbit does not completely fit the observations. It seems that scientific theories gain their non-indexicality at the expense of their applicability.

Another way of capturing the notion of indexicality is to recognise its connections to 'forms of life'. All indexical statements are embedded more or less explicitly in a form of life. Non-indexical statements attempt to transcend or deny a form of life. In the attempt to maximise the objectivity of scientific theories and to display them as universal truths they are increasingly distanced from their forms of life and consequently lose their connection to the world. This, however, does not make them either useless or powerless. They gain great strength and efficacy in the social and cultural domain. Once again this point is illustrated by maps as we shall see in Exhibits 8, 9 and 10.

Exhibit 5 ABORIGINAL-AUSTRALIAN MAPS

by Helen Watson
with the Yolngu community
at Yirrkala

Paintings by Aboriginal Australians are not immediately recognisable as maps. Nonetheless Aborigines sometimes see them as maps and so now do some Western anthropologists (Williams, Peterson and Morphy – see Further Reading list). In Exhibit 5, we shall examine whether, and in what sense, the graphic representations we conventionally call 'Aboriginal bark paintings' are maps. In particular, three barks that were presented to Deakin University by the Yolngu* community of the Laynhapuy region of NE Arnhemland will be studied. (Refer to *Singing the land, signing the land* for more information about this community. ITEM 5.1 shows the location of Yolngu homeland centres in the Laynhapuy Region of NE Arnhemland.) To avoid prejudging the issue of whether and in what way they are maps, we shall call the barks by the name that Yolngu give them, *dhulaŋ*. (We write Yolngu words using the orthography that Yolngu use. For a full description, see *Singing the land, signing the land*. The pronunciation of Yolngu words that appear in this exhibit are given in the accompanying box.) First, we must consider the conceptual framework that surrounds their production and use by Yolngu.

Preparing *dhulaŋ* is a common pastime for many Yolngu, for *dhulaŋ* serve a number of important purposes in the life of the community. *Dhulaŋ* are representations of concepts which are part of what Yolngu explain as *'djalkiri'. Djalkiri* is a generic term, often translated as 'footprints of the Ancestors'. For Yolngu to say that the notion of *djalkiri* is structurally important in Yolngu life would be akin to an English speaker saying that the notion that material things have qualities is a basic concept in her way of life. *Djalkiri* is one of the network of concepts through which time, space, personhood and community are constructed in Yolngu life.

A person's or a clan's *djalkiri* could be called their 'songline'; it refers to what English speakers have come to call 'the Dreamtime' or 'the Dreaming', the 'other time' when the people of the two great ancestor clans socialised the landscape by living in it, thus variously creating the *Dhuwa* and the *Yirritja*, the dual sub-worlds of the Yolngu world. In the course of their everyday doings the Ancestral People left their 'footprints' and 'tracks', and this *is* the now known landscape. In talking of their *djalkiri*, a speaker refers to a specific series of stories, songs, dances and graphic representations about that creative epoch, as well as to the country defined by those stories, songs, dances and graphic representations. This is the country 'owned' by that person or group. The 'whole country' is constituted by a network of tracks which intersect and define a framework for the political and economic processes of Yolngu society.

That 'other time' transcends the present; the landscape, in being a series of narratives related in specific ways, is also transcendent. The Ancestors, frequently in the form of animals (water goannas, salt water crocodiles, dugong and the like), travelled from place to place, hunted, performed ceremonies, fought and finally turned to stone or 'went into the ground', where they still remain. The actions of these powerful beings created the world as it is known today. They gave the world its forms, and its identities – its names.

Whenever we wish to refer generally to the original inhabitants of Australia, we have used the term 'Aborigine/s'; whenever possible, we use the specific term used by the people themselves, for example 'Yolngu'.

The bush is criss-crossed with their lines of travel and just as a person's or an animal's tracks are a record of what happened, the features of the landscape – hills, creeks, lakes and trees – are the record, or the story, of what happened in the Dreaming. While particular actions give name and identity to each location, the fact that together, in a certain sequence, the named places constitute journeys by particular Beings, who themselves are related in particular ways, links all identified places into a whole. It is not only the landscape that assumed its identity at this time; all things gained their identities, their places in the scheme of things.

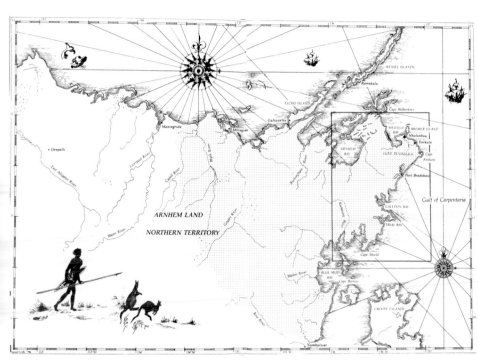

5.1
Small-scale map showing the location of Yolngu homeland centres in the Laynhapuy Region of NE Arnhemland.

Dholtji
Nadayun
Matamata
Melville Bay
Dhambaliya
Rorruwuy
Gunyaŋara
Yänuŋbi
Yirrkala
Arnhem Bay
Dhälinybuy
Guluruŋa
Yalaŋbara
Bawaka
Gurrumuru
Ŋaypinya
Garrthalala
Biranybirany
Caledon Bay
Wandawuy
Gurka'wuy
Maywundji
Bukudal
Gängan
Bäniyala
Djarrakpi

Yolngu knowledge is coincident with the creative activity of the Ancestral Beings. They traversed the land and in the process created the topography. What they did then provides the names of places along the path; the identity of each place is established by its connections to other places. Their actions also link groups of people. In turn, these links are given a social form and determine the social and political processes of Yolngu life. Thus the landscape, knowledge, story, song, graphic representation and social relations all mutually interact, forming one cohesive knowledge network. In this sense, given that knowledge and landscape structure and constitute each other, the map metaphor is entirely apposite. The landscape and knowledge are one as maps, all are constituted through spatial connectivity.

For Yolngu, what provides the connections between places—bits of socialised topography that are known through being named—are the tracks of the Ancestral Beings, and the tracks *are* the landscape. For Westerners, the connections between places are seen in terms of abstract qualities such as length or width. In a profound sense the Yolngu 'theory of land' has the landscape as a map of itself. In considering this paradox return to Exhibit 1. Yolngu resolve this paradox one way; Westerners resolve the paradox in the contrary way.

The fact that Ancestral Beings socialised the landscape and thus created its identity in that 'other time' does not mean, however, that the world is unchanging. The interrelated cosmos must be maintained by constant intervention—negotiation and renegotiation—by those responsible. There are no dualistic oppositions here, between good and bad, right and wrong, background and foreground. All elements of the world are constitutive of all other elements in the cosmos, through being related to them, and are in some sense responsible for them.

Yolngu knowledge is a commodity, or a product. You can earn it, trade it, give it and, more importantly, restrict access to it and hence use it as a means of control. It provides the basis for ceremonial power in the profoundly egalitarian Yolngu world. Moreover, the knowledge network is not transparent or passive, it is the real stuff of interaction between groups, and depends for its existence on constant activity, singing, dancing and painting. Through constant negotiation everyone knows who is responsible for what part of the knowledge network, who is charged with the care and maintenance of what song, and what land.

It seems that in raising the question of the way in which *dhulaŋ* are maps we may have gained a general insight into the construction of meaning. If the process of acquiring topographical knowledge occurs at least partly through the process of naming, then the connectivity that provides for knowledge in general is that of the network of meaning. According to this account of meaning, words acquire their meaning not simply by reference to objects but by their relationship to one another in a three-dimensional network. Thus, while the notion of a network is essential to all cultural formations, the way it manifests varies from one culture to another. In Yolngu life the knowledge network is taken as there for all to see, so long as they know what to look for; but then it takes a long time to learn that. This knowledge network is tangible and must be actively maintained.

Interested as we are in landscape and its representation, *wäŋa* too is a useful concept for us to investigate. Orientation in space is of prime concern for Yolngu. Any recounting, whether ancestral, historical or contemporary, is framed by a discussion of place: where events happened. Events coalesce in space rather than in time; landscape punctuates stories, and behind this is the 'working assumption' that human activities 'create' places by socialising space.

Yolngu do not build a domesticated space and oppose it to 'wild nature' outside this domain, instead they are truly at home when they walk through the bush, full of confidence in knowing it as their own place. A camp can be made almost anywhere within a few minutes: a smooth place to sit or lie, a fire and perhaps a billycan of tea. Previously unmarked country becomes a camp, a *wäŋa*, with all the comforts of a familiar sitting room. The way of thinking which enables people to make a camp almost anywhere they happen to be, with no sense of dislocation, is a way of thinking that has the country as its own universe of meanings. Yolngu, who like to move and shift so regularly from specific place to specific place, have made space their own. Space is socialised; landscape, a home, a *wäŋa*.

But *wäŋa* is more than this; it is in fact a very complex bundle of concepts. In calling up notions of socialised space it is used in connection with the temporary site at which people live, but it also means 'enduring country'–a named place. *Wäŋa* in this sense carries with it connections to all other named places. *Wäŋa* is not only the human creation of 'camp' but also the Dreaming creation of 'country', a concept primary to Yolngu social organisation. In the first *dhulaŋ* we shall consider (see ITEM 5.2), the salt water crocodile (*bäru*) is said to have its legs splayed out 'holding the *wäŋa* with his feet' (Gulumbu Yunupiŋu, 1987).

The interpretations of the *dhulaŋ* given here are the product of conversations between Helen Watson of Deakin University and the artist Djamika Munuŋgurr and his wife Gulumbu Yunupiŋu and daughters Ŋalawurr Munuŋgurr and Wuyuwa Munuŋgurr in April 1988. The discussions were held with the specific intention of generating the information in a form suitable for this publication.

If the *dhulaŋ* in ITEM 5.2 is oriented with the crocodile's head pointing upwards, then it can be recognised as a map in the conventional Western sense. The place mapped here is called Ŋalarrwi, near Biranybirany (the homeland centre of the Gumatj people, which can be located on ITEM 5.1). The position of the crocodile's feet and body define the area of the Gumatj clan homeland by holding on to it. The coast is represented by the rear legs, the mouth of the river is where the tail joins the body.

By looking at the detailed Western map of Caledon Bay (ITEM 5.3) you can identify the section of river that is being referred to here. The land which is owned by the Gumatj

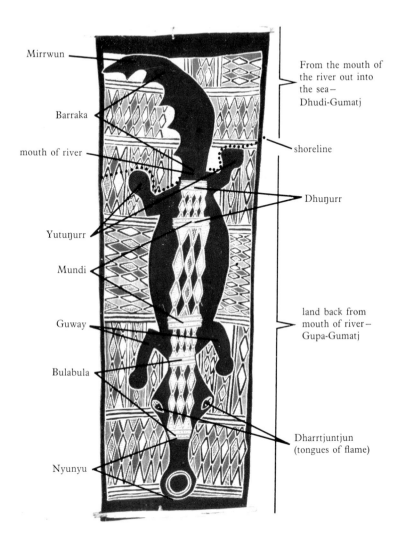

Mirrwun

Barraka

mouth of river

Yutuŋurr

Mundi

Guway

Bulabula

Nyunyu

From the mouth of
the river out into
the sea —
Dhudi-Gumatj

shoreline

Dhuŋurr

land back from
mouth of river —
Gupa-Gumatj

Dharrtjuntjun
(tongues of flame)

5.2

Crocodile and fire dreaming; moiety – *Yirritja*; clan – Gumatj; painter – Djamika
Munuŋgurr, 1985.

This *dhulaŋ* 'maps' the homeland of the Gumatj clan. The names of the parts of
the crocodile are the names of areas of land. There are names for general areas
like Gupa-Gumatj and Dhudi-Gumatj, which name separate sections of the clan.
Also there are very specific names like Djarrtjuntjun, which names a small and
restricted area.

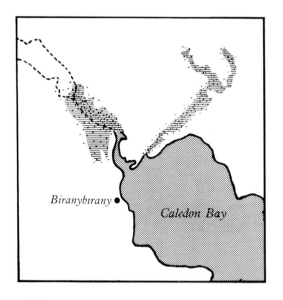

5.3
Western map of Caledon Bay.

clan extends in all directions from this river. Its boundaries are not shown here for to do so would be 'a breach of good custom' (Nancy M. Williams, *The Yolngu and their land*, 1986, p. 19).

The *dhulaŋ* represents a specific place where the crocodile (an Ancestral Being) lives, and the graphic elements are organised on spatial principles: that is, they are intended to correspond to elements of the landscape. Hence it is a map. However, it is obviously a highly conventional map. In order to able to read it, you have to know something of the stories, songs and dances of the creation of this landscape by this Ancestral Being and his kin.

The background pattern, or *mittji*, is 'the fire dreaming', a design owned specifically by the Gumatj. The irregularities of the 'diamond design' indicate flickering flames going in all directions. Here fire is a metaphor for knowledge:

> Fire always has the same elements: flames (the red diamonds); ashes (yellow diamonds); sparks (white dots); charcoal (black lines); flaming coals (white with red lines); dust (white lines on yellow) but in any particular fire these elements combine and manifest in different ways. So it is with knowledge: the general elements and their relations are always there but the way knowledge is revealed anew to each generation is particular for that generation and that time.
>
> Gulumbu Yunupiŋu, 1987

Each part of the crocodile has a name and the 'outside names'—those that can be generally known—are shown in ITEM 5.2. The pattern we see on the back of this crocodile, and indeed all crocodiles, represents the murky water which is its *wäŋa*:

> *Dhulaŋ* like this one are for teaching children—they learn from the bark and from the land itself. Children can learn the shape of the land from the bark, and from being instructed about what is written there. From being instructed they can get a map in their heads. Children can learn to have respect for the *wäŋa* in this way and the *wäyin* (game animals) that live there, and learn to mind it properly. If they don't do that it will take its revenge.
>
> Gulumbu Yunupiŋu, 1987

The bark shown in ITEM 5.4 is a large-scale portion of the one shown in ITEM 5.2: it is one side of the river. The name of the place represented here is Badaymirriwuy. If you stand at the mouth of the river facing the land, it is on the left-hand side. Beside this sandy stretch of river bank are the crocodiles' water holes. The name of that place is Buykala.

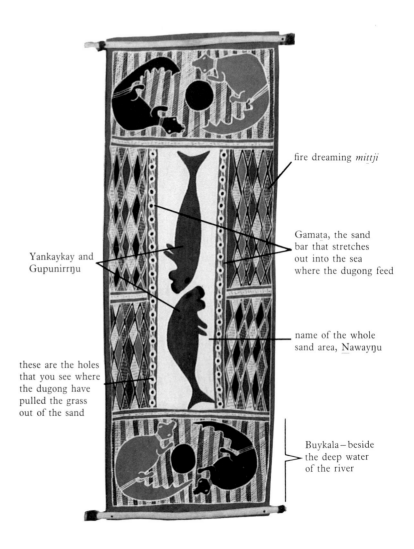

fire dreaming *mittji*

Gamata, the sand
bar that stretches
out into the sea
where the dugong feed

Yankaykay and
Gupunirrŋu

name of the whole
sand area, N̲awayŋu

these are the holes
that you see where
the dugong have
pulled the grass
out of the sand

Buykala—beside
the deep water
of the river

5.4
Dugong and fire dreaming; moiety—*Yirritja*; clan—Gumatj; painter—Djamika
Munuŋgurr, 1985.

The central white area is the sand bank N̲awayŋu. It stretches out at the mouth
of the river, the area of the crocodile's tail in ITEM **5.2**. The dugong (two
Ancestral Beings) feed on the sea grass on the edge of the sand bar Gamata. In
doing so they leave the holes depicted around the white area.

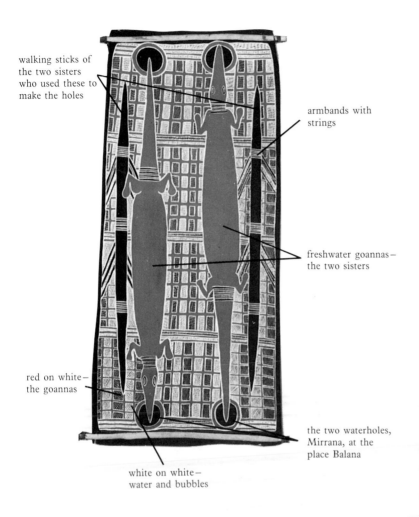

walking sticks of
the two sisters
who used these to
make the holes

armbands with
strings

freshwater goannas –
the two sisters

red on white –
the goannas

the two waterholes,
Mirrana, at the
place Balana

white on white –
water and bubbles

5.5
Water goannas and water dreaming; moiety – *Dhuwa*; clan – Djapu; painter –
Djamika Munuŋgurr, 1985.

This *dhulaŋ* portrays the episode in which a *Dhuwa* place, Balana, was created
and named by the Ancestors, the two Djan'kawu sisters. In this place, a green
leafy grove, the two sisters, the water goannas, made the water holes with their
walking sticks. The name of the water holes is Mirrana. The sisters both enter
and re-emerge from the ground here, shown by the fact that the holes are at
both ends of the animals. The background patterning, the *miṯṯji*, represents the
'water dreaming', which is a metaphor for the way knowledge continues in its
structure unchanged – it continually bubbles up in the same place. The pattern
carries the idea that, while for each generation knowledge is revealed anew, the
structure of the knowledge is always there, continually bubbling up.

The organisational principle employed in ITEM 5.5 is not spatial, it carries a narrative which is specifically located. Though it depicts topographical features we would hesitate to call it a map. As an account of what happened at one particular stopping place in the journey of the Ancestral Beings, it could be described as a diagram rather than as a map.

Only after much deliberation can we decide whether to translate *dhulaŋ* as 'painting', 'map', 'diagram' or 'graphic representation'. In making this decision, it may be useful to seek out other bark paintings to study. For example, refer to ITEM 4.40 in *Beasts and other illusions* (p. 55) and ITEM 4.3 in *Imagining landscapes* (p. 51). What do you make of these *dhulaŋ*? Can a European attempt to recount some of the information encoded here? Does it help to know something of the theory behind the production of *dhulaŋ*? In Exhibit 4 in *Imagining landscapes* (ITEMS 4.4, 4.5, 4.7 and 4.15) we see maps produced by Aboriginal people with a different set of graphic conventions. One intuits that these mapmakers share something of the Yolngu 'theory of the land' from the explanations that are provided here.

In considering the ways in which *dhulaŋ* are maps, we must not lose sight of the fact that *dhulaŋ* are also religious icons. They have at least two aspects in common with Christian religious icons. Firstly, they present another, 'transcendental', world; they carry with them the presence of *Waŋarr*–the Time and the Beings of the Dreaming–in much the same way Christian icons present 'the living God'. The second element they have in common with Christian icons is that they are a 'gloss on text', the narrative of the Dreaming. *Dhulaŋ* are maps only insofar as the landscape is itself a 'text'. Unlike Western maps, *dhulaŋ* do not seek to represent the text of landscape 'in ratio'; they do not seek the 'rational representation of nature'. *Dhulaŋ* and Western maps have different theories of picturing because they are produced within different theories of knowledge. It is because the Yolngu knowledge system has landscape itself as a meaning system, through which meanings are made following the actions of the Ancestral Beings, that *dhulaŋ* are coincidentally maps and icons. Rights in graphic representation, in *dhulaŋ* and other ways of presenting the Ancestral World (song and dance) are inextricably interwoven. The Ancestral Beings handed over the land to *particular* human groups. The condition of maintaining the world in this proper state is that those groups continue to perform ceremonies and produce the paintings and ceremonial objects that commemorate their creative acts.

We might say that in a profound way the Ancestral Beings of the Yolngu were mapmakers. They created the landscape and at the same time made the country a map of itself in the knowledge network. They created the symbols and the ways of their use so that the map might be read by those to whom these things have been revealed. Reading the map is penetrating deeper into the texture of the knowledge network–the land itself.

Pronunciation of Yolngu words	
Balana	Ba-la-na
Badaymirriwuy	Bud-ae-mirri-woi
bäru	bar-roo
Biranybirany	Bir-ane-bir-ane
Buykala	Boi-kala
dhulaŋ	dthoo-lung
Dhuwa	Dthoo-wa
djalkiri	djal-kiri
Djan'kawu	Djan-ka-woo
Gamata	Ga-ma-ta
Gumatj	Goom-ach
Mirrana	Mirr-a-na
miṯtji	mint-jee
Ŋalarrwi	Ng-a-larr-wi
Ṉawayŋu	Na-wae-ngu
wäŋa	waa-ng-a
Waŋarr	Wa-ng-arr
wäyin	waa-yin
Yirritja	Yirr-i-cha
Yolngu	Yol-ng-oo

Exhibit 6
THE STORY
SO FAR

In this exhibit we turn to a modern topographical map to see how the issues we have raised so far bear up under examination. Closely examine the Ordnance Survey map (ITEM 6.1) and the two photographs (ITEMS 6.2 and 6.3) of the landscape depicted. Consider carefully the following questions.

- How do you know ITEM 6.1 is a map?
- How do you read it? Do you need anything that is not on the map to read it?
- Why is it called an Ordnance Survey map?
- What kind of symbols does it employ? Are they purely arbitrary, symbolic, natural, iconic? Are they all explained in the key? On what kind of conventions are they dependent? Can anyone read the map or is there tacit knowledge involved?
- Is there anything indexical about the map?
- Is there anything invariant between the map and reality, that is to say non-indexical?
- Is a scale necessary? What does it enable you to do?
- Does the map contain more information than is specifically recorded by the cartographer?
- Is there any information lost?
- Does it contain all the possible information? Can there be multiple maps of the same place?
- What sort of projection is used? Are you told? What difference does it make? Are projections necessarily conventional?
- What orientation does it have? Why is north usually at the top? Does it matter?
- What are the criteria by which one evaluates a map? Are they all necessarily tied to human purposes? Does 'workability' cover them all?
- What possible uses are there for maps? Are any maps anything else as well as maps?
- What does the grid enable you to do? If one of the functions of the map is to allow accurate measurements to be made, what sort of things can be measured? What social accomplishments and practices are required to enable those measurements (bench marks, sea level, National Grid datum, standard yard etc.)?
- How many of the features in the map are non-geographical, that is, constructed? How many of them depend on continuing social and political practices for their existence?
- Are any of the functions of the society, state, interest groups, military served by this, or any, map?
- What is the difference between a map and a photograph? Which is more real?
- The two aerial photographs are of the top right-hand and bottom left-hand quarters respectively. Did you find any difficulty in matching them? What does this tell you about the representational power of the map?
- What do the photographs show that the map does not? What are the differences between them with respect to time? What is the difference in function? Why can't we replace maps with photographs?

6.1
An Ordnance Survey map,
scale 1:63 360, of the city of
Canterbury, published in 1959.

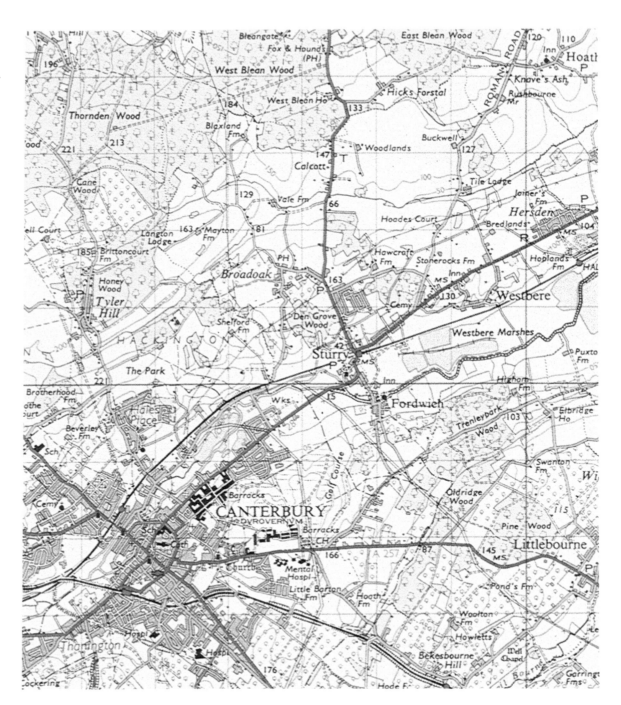

6.2
Aerial photograph of top right quarter of ITEM **6.1**, taken in 1967.

6.3
Aerial photograph of bottom left quarter of ITEM **6.1**, taken in 1967, showing a new road on the west side of Canterbury.

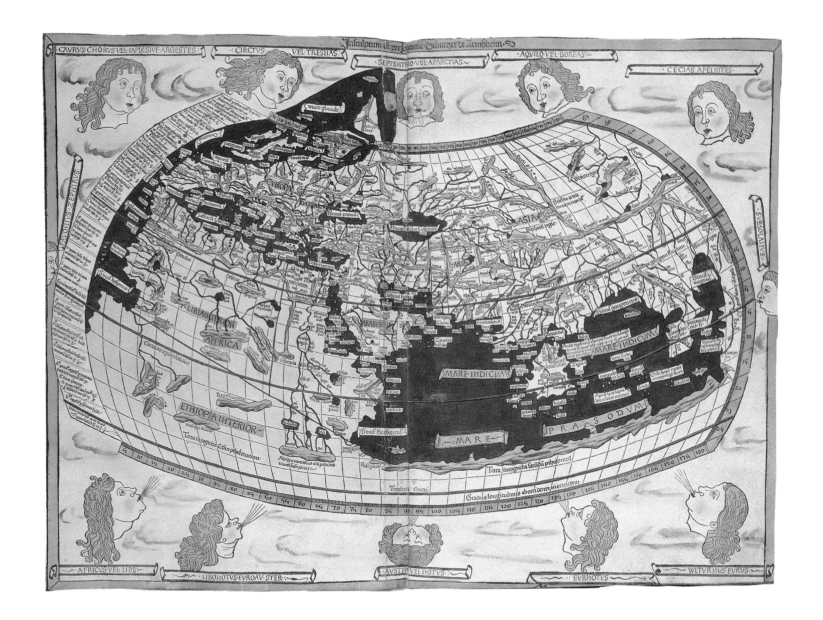

Exhibit 7
THE FUNCTION
OF MAPS

We earlier referred to the tendency to assume that the acme of mapmaking is the large-scale Western topographic map because it is seen to be so 'accurate'. But can accuracy be assessed independently of function? After all an Ordnance Survey map can be said to be inaccurate to the degree to which it omits details or, for example, does not use a standardised representation for equivalent items. In the Ordnance Survey map of Exhibit 6 (ITEM **6.1**), you can see that some country pubs are given a name while no city pubs are. Does this mean that country pubs are bigger or that there are no pubs in Canterbury? No, obviously the reasonable response to such criticisms is that the accuracy can only be assessed in the light of the purposes for which the map was intended. As E. H. Gombrich has pointed out, a pictorial representation

> is not a faithful record of a visual experience, but the faithful construction of a relational model . . . Such a model can be constructed to any required degree of accuracy. What is decisive here is clearly the word 'required'. The form of a representation cannot be divorced from its purpose and the requirements of the society in which the given visual language gains currency.
>
> E. H. Gombrich, *Art and illusion*, 1960, p. 90

If it is true that accuracy is linked to function, then indexicality cannot be simply equated with mere practicality. In other words, highly accurate maps are not really less indexical or less tied to their context of use. Instead it might be more illuminating to see various maps as having different modes of transcending indexicality.

As we have seen in the case of the Ordnance Survey map, it cannot simply be 'read', even in conjunction with the key. Contour lines, for example, make no sense at all without training and facility in imagining a three-dimensional projection. Equally, contour lines can have no meaning without the existence of a plethora of assumptions, institutional arrangements and systems of measurement. The concept of sea level, the infra-structure of benchmarks, national surveys and instruments such as theodolites all constitute the 'forms of life' that are taken for granted in the creation of contour lines. Indeed the state has to put in enormous amounts of money and work to maintain this system of measurement. Were it to cease to do so, the forms of life that make the map possible would collapse and this fine creation would suffer the same fate as the map in Borges's story (ITEM **1.1**). Thus, while Ordnance Survey maps are deliberately made to appear non-indexical, and to some extent are successful in that they do enable people strange to a particular piece of territory to find their way around, they are nonetheless highly indexical in that they are completely dependent on forms of life.

Let us return now to ITEMS **2.7** and **2.8** and reconsider the questions raised about them. A stranger to the culture that produced such maps as these, or the Ordnance Survey map, would be unable to read them without the appropriate training, just as Westerners cannot read aboriginal maps, or even their own maps, without training. Thus the claim by Westerners that aboriginal maps are more indexical and hence less scientific than

◀ **7.1**
The first woodcut world map, following Claudius Ptolemy, the 2nd-century Alexandrian geographer and astronomer, with the contemporary addition of Scandinavia and Greeenland (1482). Note the grid, and the the southern 'Terra Incognita'. Once it was possible to conceive of the whole world as a map, Greek aesthetic sensibilities dictated that there should be a southern body of land to balance that in the north. Australia was thus invented through the power of the map before it was 'discovered'.

The term 'Aborigine' (upper-case 'a') refers specifically to the original peoples of Australia, whereas the term 'aborigine' (lower-case 'a') is used generically to refer to indigenous peoples around the world.

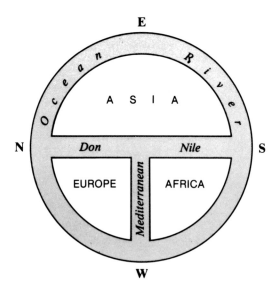

7.2
Early Western Christian *mappamundi* (maps of the world) are called T-O maps. The world is divided by the three major waterways symbolising the cross, with east at the the top and Jerusalem at the centre.

theirs springs largely from the transparency of the forms of life in which their maps are embedded. This reflects a difference in the ways the differing cultures achieve a transcendence of indexicality, rather than a difference in their correspondence to reality.

Aboriginal maps can only be properly read or understood by the initiated, since some of the information they contain is secret. This secrecy concerns the ways in which the map is linked to the whole body of knowledge that constitutes Aboriginal culture. For Aborigines, the acquisition of that knowledge is a slow ritualised process of becoming initiated in the power–knowledge network, essentially a process open only to those who have passed through the earlier stages. By contrast, the Western knowledge system has the appearance of being open to all, in that nothing is secret. Hence all the objects on the map are located with respect to an absolute co-ordinate system supposedly outside the limits of our culture.

One could argue that in Western society knowledge gains its power through denying, or rendering transparent, the inherent indexicality of all statements or knowledge claims. In the Western tradition the way to imbue a claim with authority is to attempt to eradicate all signs of its local, contingent, social and individual production. Australian Aborigines on the other hand ensure that their knowledge claims carry authority by so emphasising their indexicality that only the initiated can go beyond the surface appearance of local contingency.

In the light of these considerations we should perhaps recognise that all maps, and indeed all representations, can be related to experience and that instead of rating them in terms of accuracy or scientificity we should consider only their 'workability'–how successful are they in achieving the aims for which they were drawn–and what is their range of application.

If you look at ITEMS 7.1, 7.2 and 7.3, some rather curious features of the development of Western maps emerge. Ptolemy first drew his map in the 2nd century AD and it displays an impressive knowledge of the world. It contrasts strongly with Hereford map from 1300 AD, which by comparison is highly inaccurate and limited in its coverage. However, the Hereford map is a T-O maps (ITEM 7.2) and serves religious interests as well as topographical ones. The orthodox account would have it that Western maps progressed; they became more accurate and more scientific. The apparent lack of progress shown in these maps is better understood if the differing interests of the makers are taken into account. Thus world maps serve different functions from topographical maps and have to be evaluated accordingly.

The Mexican land claim map (ITEM 7.4 and see also back cover) and the English and Roman route maps (ITEM 7.5 and front cover) have all the appearances of being distorted, conventional and totally indexical. However, once recognised as route maps their

7.3
The Ebstorf map (1235) is an example of
mappamundi, drawn on T-O principles despite the
availability of Ptolemaic maps like ITEM **7.1**. Note
Christ's head, hands and feet around the edge of the
map.

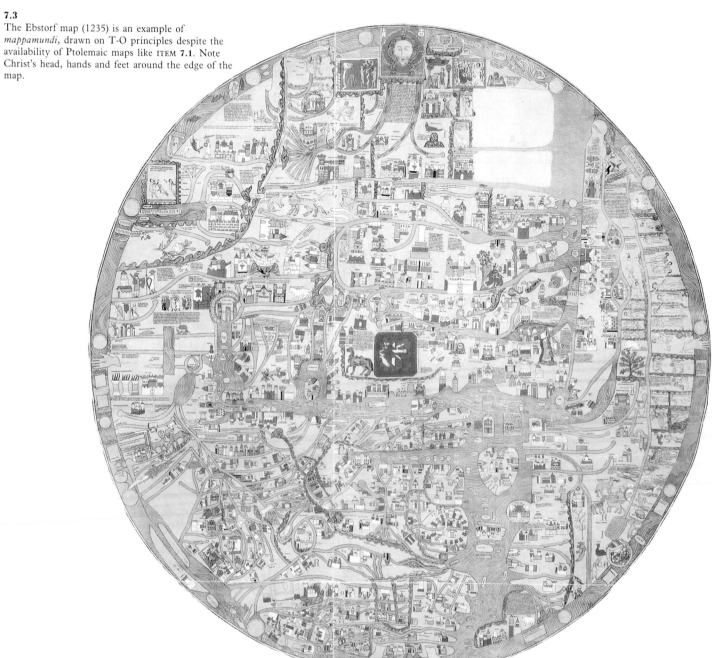

workability becomes apparent. They embody a lot of accurate detail and information. Apart from the language and symbols which require translation, that is, a key, they are as readable as the London Underground map (ITEM 2.8). Their base in a common human purpose in fact makes them more readable than Sagan's plaque (ITEM 2.7), for example.

Maps can have a variety of functions: they can make political jokes (ITEM 7.6); they can educate and entertain (ITEM 7.7); and they can tell lies (ITEM 7.8). All maps also have a latent symbolic function, for example, legitimating and disseminating the state's view of reality.

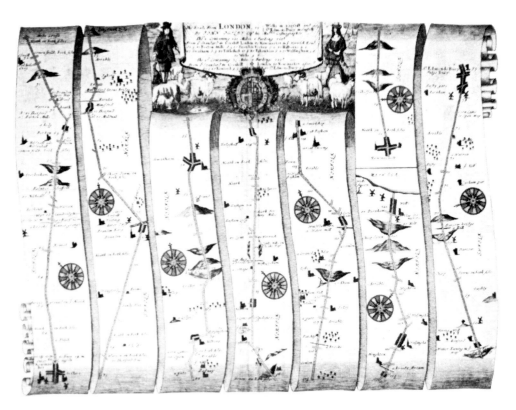

7.4
Early 16th-century Mexican Indian map of Metlacoyuca setting out lordship of the area. The temple marks the town and the figure, the lord's genealogy.

7.5
'The Roads from London to Wells in Norfolk and St Edmons Bury in Suffolk', a strip map (1670). A highly conventional route map (note the reversed hills to indicate that the traveller is going downhill), but just as effective as the Automobile Club route maps of today.

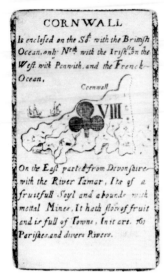

7.6
Cartoon map.

7.7
These playing card maps serve to educate as well as
to entertain. The upper pair are by Hoffman (1678)
and the lower pair by Redmayne (1711).

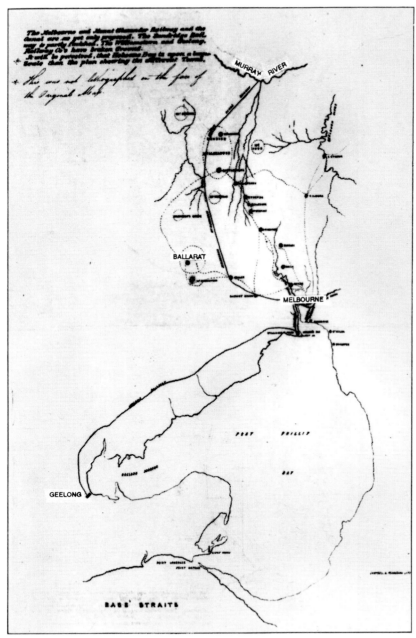

7.8
The 'false' goldfields map with Strachan & Co's
explanation and the 'real' map.

The false map

The importance of a locality or township in Victoria is increased in proportion as it is well or badly situated relatively to the Gold Fields; and facility of approach, and shortness of distance, confer claims, and offer advantages, the value of which is fully appreciated, as well in the old country as in this. The position of Melbourne and Geelong, in this respect, has been made the subject of controversy by the public press; and that a jealousy on this head has arisen is too apparent to be overlooked, and is readily accounted for when the vast natural advantages of Geelong are candidly viewed. The progressive pace at which this town and its commercial interests have been moving onwards must satisfy its most ardent well-wishers, notwithstanding the obstructions wilfully thrown in its way. We do not hesitate to affirm that there have been few places in which a greater amount of sound business, proportionately, has been done within the last three years, than in Geelong. The business now carried on here may be learned at once, when it is stated that the imports for the year ending 30th June, 1854, amounted to £2,211,571, and employed 114,656 tons of shipping; and that the exports during the same period amounted to £1,316,813.

The fact, however, that this jealousy does exist has been recently shown most flagrantly by the publication of a MAP purporting to be a correct representation of the distances of Melbourne and Geelong from the Gold Fields of this Colony. It is attached to a general Price Current, issued by a firm in Melbourne, and is evidently intended to be circulated throughout the world. Its palpable object is to show that the Gold Fields and a highly populated district are in close proximity to Melbourne, while Geelong is placed at a false distance. Ballarat is made to appear just twice the distance from Geelong that it is from Melbourne, whereas the real distance of that township and its Gold Fields is only forty-eight miles, in a straight line from Geelong, and, by the same mode of measurement, sixty-two miles distant from Melbourne; and the same glaring disproportions exist with regard to Mount Alexander, the Murray River, etc. Who the concoctors of this disreputable document were we know not, for it bears no signature—but, for the purpose of indentification, we state that the lithographers were Messrs. Campbell and Fergusson, Melbourne.

In order the more fully to exhibit the baseness of this fraud, a map is enclosed, published by Messrs. Macdonald & Garrard of Geelong, on a correct scale, and a moment's comparison of the False Map with the true one will point out the dishonorable nature of the SILENT LIE it was the mission of this Map to tell!

Strachan & Co.
Geelong, 1st July, 1854.

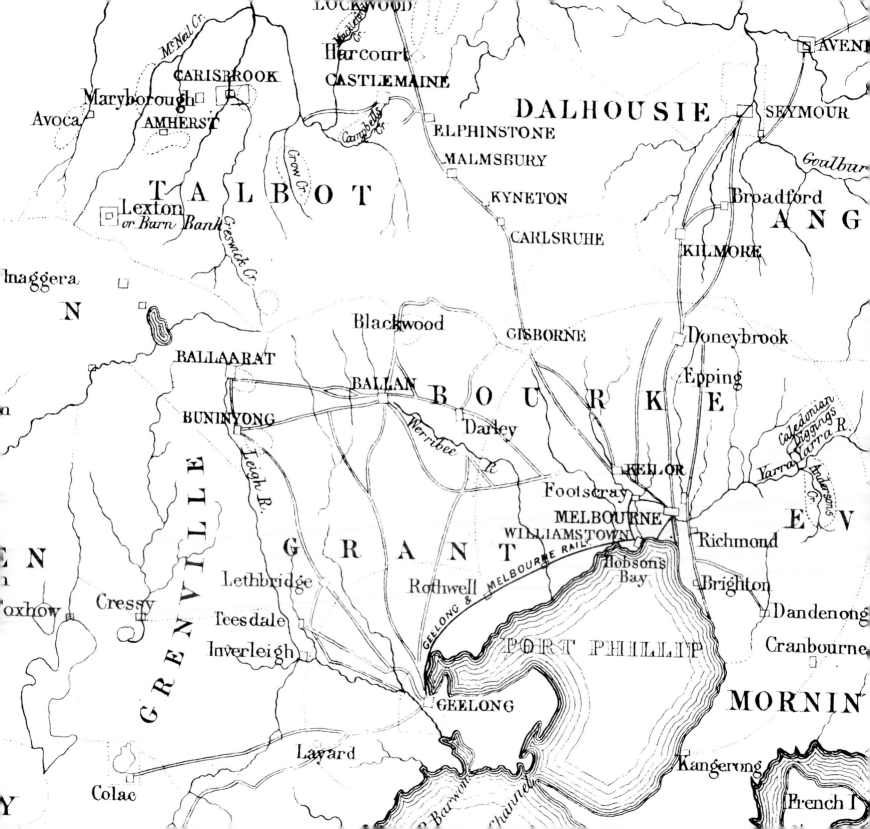

Exhibit 8
MAPS–A WAY
OF ORDERING
KNOWLEDGE

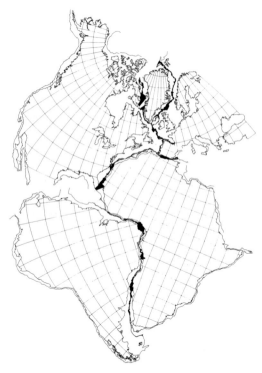

8.1
Bullard's computer reconstruction of the fit of the continents as evidence of their original connection according to Wegener's hypothesis. He used 'objective' mathematical methods to obviate the ealier criticisms of Wegener's fit. The continents are here defined by the 500 fathom contour along the edge of the continental shelves. The black shading represents overlap, and the grey shading, gaps. Latitude and longitude lines indicate present orientations of the continents.

Karl Popper argues that languages have a descriptive function which is clearly distinguishable from their argumentative function. This, according to Popper,

> makes the familiar analogy between maps and scientific theories a particularly unfortunate one. Theories are essentially argumentative systems of statements: their main point is that they explain deductively. Maps are non-argumentative. Of course every theory is also descriptive, like a map.
>
> Karl Popper, *Unended quest: an intellectual autobiography*, 1976, p. 77)

Popper may be mistaken about maps, possibly because he is concerned at this point to make a distinction between the descriptive and argumentative functions of language. Though the distinction can be sustained analytically, it cannot be pushed too far because there is a powerful sense in which descriptions *are* arguments. The strongest sense in which that is true is illustrated by maps. They invariably carry less information about the environment than is out there, since they are necessarily selective, but they also frequently carry more information than was actually recorded. Some maps, like those used for navigation, can have data plotted on them and deductions as to position and distance to destinations made from them. But the example of Wegener's theory of continental drift (see ITEM 8.1) and the cholera map (ITEM 8.2), show that new knowledge can be gained from a map in a profound and significant way. It was Wegener's 'fitting' of the continents that led him to hypothesise the original joining of South Africa and South America. It is interesting that, while Wegener used that 'fit' as evidence for his theory, his opponents were able to reverse the argument and criticise his theory on the grounds that map-fitting did not constitute evidence and that the apparent fit was the result of selective distortion. Mapping the outbreaks of cholera also revealed a distribution that could not have been simply read from the data. Moreover maps are not always mere re-presentations of data. In ITEM 8.3 the magnetic stripes in rocks on the sea floor off the coast of West Canada become in their turn evidence of the phenomena of sea-floor spreading from ocean ridges, the ultimate proof of Wegener's theory. Indeed ordering information spatially provides a very powerful mode of inference to new knowledge that is ignored by Popper.

However, there is an even deeper sense in which maps are like theories in that they are 'argumentative systems of statements'. That is the sense in which they embody or express a cognitive schema. As has been discussed in *Imagining nature*, observation statements are not clearly separable from theoretical statements, and theoretical statements in turn embody sets of assumptions about how reality is ordered. This whole complex of unconscious assumptions about the ordering of reality which structures our experience of it can be thought of as a cognitive schema. In the case of maps, the idea that our ability to understand the world is dependent on modes of ordering of which we are at best only partially aware is of particular interest. As concrete examples they provide an opportunity for bringing such cognitive schemas to the fore, and they also provide an opportunity to explore the claim that in the deepest possible way knowledge is inherently spatial, and embedded in practical action.

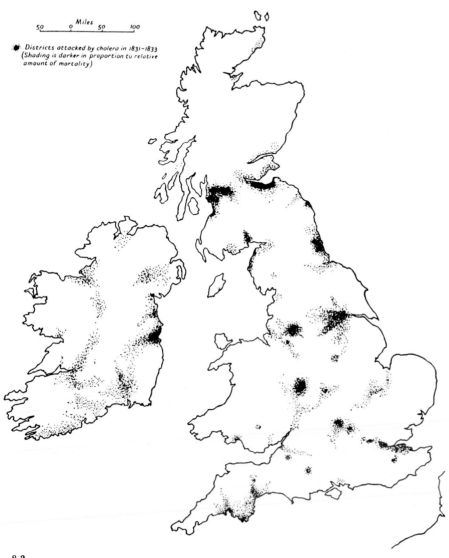

8.2
A 20th-century redrawing of Augustus Petermann's map of 1852, displaying the spatial impact of the cholera outbreak of 1831–33.

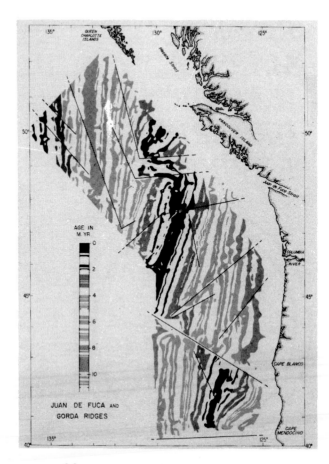

8.3
The mapping of the magnetism of rocks beneath the sea reveals mirror image patterning on either side of the central ridge, which is graphic evidence of molten rock flowing out from the ridge and thereby accounting for sea-floor spreading off West Canada.

Exhibit 9
MAPS–A
WAY OF
ORDERING OUR
ENVIRONMENT

9.1
Compass rose

Just as maps can provide us with new knowledge by ordering it spatially, so do they provide ways of ordering and knowing our physical environment – the territory. Consider again a passage from Exhibit 1:

> As we experience space, and construct representations of it, we know that it will be continuous. Everything is somewhere, and no matter what other characteristics objects do not share, they *always* share relative location, that is, spatiality; hence the desirability of equating knowledge with space, an intellectual space. This assures an organization and a basis for predictability, which are shared by absolutely everyone. This proposition appears to be so fundamental that apparently it is simply adopted a priori.
>
> A. H. Robinson & B. B. Petchenik, *The nature of maps: essays towards understanding maps and mapping*, 1976, p. 4

An essential way to look at what is implied in concepts like Robinson and Petchenik's 'spatiality' or Lewis's 'connectivity' is to explore their relationship to doing things in the material world – to consider how it is that we navigate or find our way about.

It might be thought that the ultimate evidence of the superiority of Western maps is their use in navigating on unknown waters or in unknown territory. If they are informed by a theory of projective geometry that takes account of the transformations involved in presenting a three-dimensional surface on a two-dimensional one, if they use a highly systematic representational mode, if they are drawn with great accuracy, then one can find one's way across otherwise featureless or foreign terrain. Conventional wisdom has it that 'primitives' do not have maps in the proper sense since they are *familiar* with the territory that they invariably traverse with great skill. It is sometimes said that such maps as they may have are just ritual objects, aides-memoire, messages of some sort or records of past events. Such views take no account of how we actually navigate today, how we navigated before modern maps were invented or how so-called 'primitive' maps are read or used.

The compass rose (ITEM 9.1), which appears on so many maps, is now often seen as a mere decorative space filler, a hangover from the days when maps were more like illuminated manuscripts than communication devices. In fact, the compass rose 'is a very abstract model, a cognitive schema, of the relations of direction to time, of solar time to lunar time, and of time to tide. [To use a Micronesian term,] it is an *etak* of medieval navigation' (p. 266). It enabled medieval sailors to navigate successfully without literacy, writing, sophisticated instruments, the scientific method or Western schooling. They managed to negotiate the coastal waters of Europe and eventually Africa and the rest of the world without having either a map or foreknowledge. They achieved this by having a thorough understanding of the cognitive schema (C. O. Frake, 'Cognitive maps of time and tide among medieval seafarers', 1985, pp. 254–70).

To predict the tides requires a theory of the tides, a method of determining, recording and correlating solar and lunar time, and a memory of the lunar tidal schedule (the establishment of the port) for every locality. Piaget himself could not have designed a better task for testing formal operational thinking. The medieval sailor met this test ingeniously by appropriating a cognitive schema for spatial orientation – the compass rose – as an abstract device for recording and calculating time and tide.

★　★　★

He who wishes to learn to calculate the tides must first know all the points of the compass with its quarter points and half points, since this is the essential foundation of this matter and without it there can be no certainty.

Portuguese sailing
directions, c. AD 1500

(Charles O. Frake, 'Cognitive maps of time and tide among medieval seafarers', 1985, p. 262)

In order to find our way successfully, it is not enough just to have a map. We need a cognitive schema, as well as practical mastery of way-finding, to be able to generate an indexical image of the territory. ITEM 9.2 is a map of Juan Fernandos Island. To identify the island, you have to generate an image of what the island looks like from your position, as in the landfall sketch at the top. Thus indexical images are required in addition to the supposedly non-indexical information on the map (A. Gell, 'How to read a map: remarks on the practical logic of navigation', 1985, pp. 271–86). Though having a map makes the task of navigation a lot easier, it is not essential if you have a cognitive schema and practical mastery.

If we recall the Aboriginal bark paintings in Exhibit 5, we find it extremely difficult to see any topographical representation in them. The key map characteristic of spatiality or connectivity seems absent or sacrificed to the interests of symmetry or aesthetics. Reading them requires the acquisition of a large body of esoteric knowledge. However, David Lewis's example of the explanation given of a map of a journey he took with some Australian Aborigines shows that relative spatial location is indeed preserved. (Read ITEM 9.3.) It is accomplished through the telling of a myth that 'connects' the salient features of the landscape in the way the travellers experienced them. In other words, the bark painting can be read as a map only if you have a thorough understanding of the forms of life of Aboriginal culture. Likewise, European maps are not autonomous. They can only be read through the myths that Europeans tell about their relationship to the land.

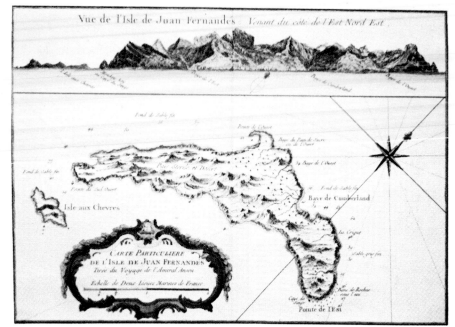

9.2
Map of Juan Fernandos Island, with a view of the island as seen from an east-north-easterly direction at the top (1754). Made by J. Bellin after the voyage of Admiral Anson.

9.3

The way of the nomad

How do Aborigines find their way across the arid
wastes of central Australia? Certainly, they achieve
feats of unerring direction finding and tracking. To
find out if their methods were in any way similar to
the non-instrumental sea navigation of the Pacific
Islanders, I travelled the fringes of the Simpson
Desert in 1972 with the Antikarinya tracker
Wintinna Mick . . . Then, in 1973 and 1974 the
investigation was extended into the Western Desert
under the tutelage of men, most of whom had spent
their youth as nomadic stone age hunters.

What I learnt from them was mainly through
practical demonstration on the trail. The lifting
above the horizon of a particular escarpment exactly
at the point predicted; our arrival at a stated
destination in accordance with intricate instructions
after hard travelling; and our finding aimed-for
waterholes – the ultimate physical necessity –
confirmed the value of their concepts and showed
that I had grasped them correctly.

The total distance I travelled with Aborigines in
those three years was seven thousand eight hundred
kilometres, including a thousand kilometres over
completely trackless terrain. Most was by Land
Rover, with a small proportion on foot. Several
journeys required the laying of petrol depots. The
longest was a round trip of sixteen hundred and
sixty kilometres from Yayayi settlement, three
hundred kilometres west of Alice Springs, to the
abandoned wells on the old Canning Stock Route,
eight hundred kilometres west of the last human
habitation in the Northern Territory.

All my preconceived ideas about 'land navigation'
turned out to be wrong. In place of the stars, sun,
winds and waves that guide Pacific Island canoemen,
the main references of the Aborigines proved to be
the meandering tracks of the ancestral Dreamtime
beings that form a network over the whole Western
Desert.

. . .

The importance of sacred sites was brought out
when I asked Aborigines to point out the direction
of distant places. In thirty-three instances, accuracy
varied between exactly right and 67 per cent error –
an enormous scatter. But every one of the big errors
occurred when non-sacred places were being

indicated. For the six far-away sacred sites the error
was never more than 10 per cent and, in fact,
averaged 2.8°.

The place was between Warren Creek and Ilpili,
in western Loritja-Aranda country, familiar to
Jeffrey Tjangala and Yapa Yapa Tjangala, both
Pintupi. It was featureless and flat, with moderately
open mulga and spear grass, devoid of sandhills,
creek beds, escarpments, tall trees or other
references. Visibility through the evenly-spaced
mulga was a hundred metres at most.

A kangaroo, wounded by a bullet, was hunted on
foot for half an hour. After it was killed, Jeffrey and
Yapa Yapa headed without hesitation directly back
towards the Land Rover that had been invisible
since the first minutes of the chase.

Q. 'How do you know we are heading straight
    towards the Land Rover?'
A. Jeffrey taps his forehead. '*Malu* (kangaroo) swing
    round this way, then this', indicating with
    sweeps of his arm: 'We take short cut.'
Q. 'Are you using the sun?'
A. 'No.'

The Land Rover duly appeared ahead through the
mulga in about a quarter of an hour. Jeffrey then
repeated his explanation, illustrating by gestures and
by drawing in the sand the *malu's* track and our own
'short cut' home.

Had Jeffrey any points of reference? The only
external one was the starting point, and the sun was
not consulted. He was not using the 'points of the
compass' (which could have been found from the
sun) nor, in this case, did sacred sites come into the
picture. It would appear that Jeffrey was orienting
on some kind of dynamic 'mental map', which was
continually being up-dated in terms of time, distance
and bearing, and more radically realigned at each
major change of direction, so that the hunters
remained at all times aware of the precise direction
of their starting point.

. . .

Next day we drove on westward over country
unfamiliar to both men. About a hundred and
seventy kilometres on, and uncertain of our exact
position, we camped in a mulga clump on gently
undulating but remarkably featureless *rira*, or stony
desert country. Jeffrey drew a cross in the sand to
represent north, south, east and west: I confirmed by
the stars that his directions were accurate. He said:
'North, south, east and west are *like this in my head.*
I know them because we were travelling west and
circled back south until we were heading south-east

when we made camp.' He denied looking at the
stars, and this was later dramatically confirmed by
his inability to drive by the stars. Jeffrey went on to
point out the directions of important Dreaming sites
all around the horizon, beginning with one near
Lake Disappointment, four hundred kilometres
away.

In similar circumstances, Billy Stockman
Tjapaltjari, and Anmatjara, explained that he always
knew the directions while we were traversing a
semicircular route 'by keeping them in my head
when we turn another way.' All these explanations
of orienting were, plainly, in terms of continually
up-dated mental maps.

When an Aboriginal depicts a stretch of country
he generally incorporates its mythical with its
physical features, so stressing the inseparable
interrelation between the two. Such paintings cannot
be interpreted without inside knowledge, yet their
emphasis on the spiritual attributes of places makes
them doubly memorable to the initiated. If such an
abstraction seems strange, it is well to remember
that Western maps, too, are often stylised. Neither
contour lines nor the soundings on a chart are
physical realities. Again, the map-diagrams on
Sydney suburban trains are quite as abstract as
anything drawn by Aborigines.

Let us now analyse a picture of Muranji
Escarpment and Dreaming. This was painted in
1974 by Big Peter Tjupurrula, a Loritja and my
kuta (elder brother), between our two visits to
Muranji. It can be 'read' only by realising these
three things: that the main Dreaming site is placed
at the centre; the rock holes and other features are
in the correct travel sequence and the corners of the
escarpment (3 and 5) are approximately correctly
oriented; the symbols representing rock holes and
the rest have to be indentified by the 'map' maker
himself.

The picture was taken with us on our second visit
to Muranji to be explained and 'sung'. It is both the
story of the Dreamtime old woman who chases a
little boy and also a 'map'. The 'singing' of this
picture was a secret-sacred rite during which a white
woman was sent out of earshot. It was also
indispensable to the 'map's' interpretation.

. . .

Here is an example of faultless orientation by the
most unobtrusive landmarks to physically or
spiritually significant places (two waterholes and a
flint knife deposit; two sacred places). The terrain
was a spinifex-covered plain, with some low
sandhills and a few undulations hardly deserving the

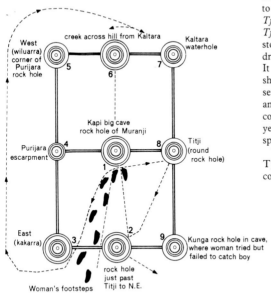

West (wiluarra) corner of Purijara rock hole

creek across hill from Kaltara

Kaltara waterhole

5   6   7

Kapi big cave rock hole of Muranji

Purijara escarpment   4   8   Titji (round rock hole)

1

East (kakarra)   3   2   9   Kunga rock hole in cave, where woman tried but failed to catch boy

rock hole just past Titji to N.E.

Woman's footsteps

The diagram (top) shows how Big Peter Tjupurrula's picture represented Muranji Escarpment and Dreaming. Below is a sketch map of the same place—'The map diagrams on Sydney suburban trains are quite as abstract as anything drawn by Aborigines'.

name of hills; it was known personally or by repute to all the nine Pintupi in the party. Our objective, *Tjulyurnya*, is a termination of the *Wati Kutjura Tjukurpa*, where a pattern of triangular yellow stones represent the Dreamtime *papa* (dingoes) that drove the two lizard men underground at this spot. It is extremely sacred and had not previously been shown to white men. The Pintupi, whose present settlement is at Yayayi, had travelled four hundred and seventy-five kilometres westward into their old country to bring back to Yayayi some of the sacred yellow stones, and to gather *mulyarti* wood for spears.

The distance from our camp at Yunala to Tjulyurnya was forty-three kilometres, all cross-country. The courses were as follows:

1. Seven kilometres a little south of west to Namurunya Soak, a tiny hollow that seemed, to my eyes, to have no identifying marks at all.
2. Thirteen kilometres south-west to a site where *kante*, sharp flints used for stone knives, were found. This was beside a low rise in the ground.
3. Five kilometres south-east, then round the end of a sandhill to Rungkaratjunku, a sacred site.
4. Winding in and out between low sandhills, generally west-south-west, sixteen kilometres to Tjulyurnya rock hole by a little hill. The sacred place was two kilometres further on.

The Pintupi's route-finding by these unremarkable landmarks was uncannily accurate. They always knew just where they were, they knew the direction of spiritually important places for hundreds of kilometres around, and were oriented in compass terms.

. . .

Aborigines travelling on foot at night take pains not to impair their night vision and hence their ability to follow the terrain. According to Yapa Yapa Tjangala, it was invariably the practice in such circumstances to have the fire sticks carried at the *rear* of the column.

. . .

Aborigines who move out from their extensive home ground into less familiar country make use of a variety of guidelines. Detailed knowledge of the myth-reinforced geography of the Dreamtime tracks, with their 'camps' and sacred sites—all real places whose physical features are often mentioned in the songs and stories—greatly expands their terrestrial horizons, but only in rather general terms. Mythical information is rarely exact enough to enable a complete stranger to locate with confidence the waterholes upon which life depends.

Animated discussion of every conceivable aspect of places visited or known by repute makes up a good part of camp and wayside conversation. This is an important factor in extending a person's range—and it solves the water problem . . .

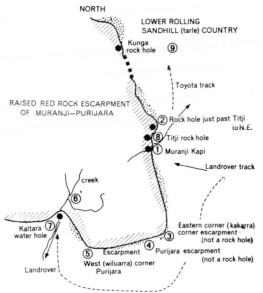

NORTH

LOWER ROLLING SANDHILL (tarle) COUNTRY

Kunga rock hole   9

Toyota track

RAISED RED ROCK ESCARPMENT OF MURANJI–PURIJARA

2 Rock hole just past Titji to N.E.

8 Titji rock hole

1 Muranji Kapi

Landrover track

creek

6

Kaltara water hole   7

Eastern corner (kakarra) corner escarpment (not a rock hole)

3

5 Escarpment   4 Purijara escarpment (not a rock hole)

West (wiluarra) corner Purijara

Landrover

(David Lewis, 'The way of the nomad', in *From earlier fleets: Hemisphere—an Aboriginal anthology*, 1978, pp. 78–82)

Exhibit 10
MAPS AND
POWER

We have found in the previous exhibits that Western maps and aboriginal maps are more fruitfully compared in terms of their range and degree of workability or usability, rather than their accuracy. However, this leaves unexamined the dimension of power. Documents, texts, diagrams, lists, maps ('discourses' in general) embody power in a variety of ways. Discourses set the agenda of what kind of questions can be asked, what kind of answers are 'possible', and equally what kind of questions and answers are 'impossible' within that particular discourse or text. In terms of maps, for example, looking at a Mercator projection you can read off the relative direction of Anchorage from London as straight line, but you cannot read off the shortest route, it being a segment of a great circle. Maps, like theories, have power in virtue of introducing modes of manipulation and control that are not possible without them. They become evidence of reality in themselves and can only be challenged through the production of other maps or theories.

Joseph Rouse argues that

> all interpretation (which includes all intentional behavior, not just discourse) presupposes a configuration or field of practices, equipment, social roles, and purposes that sustains the intelligibility both of our interpretive possibilities and of the various other things that show up within that field . . . Power has to do with the ways interpretations within the field reshape the field itself and thus reshape and constrain agents and their possible actions. Thus to say that a practice involves power relations, has effects of power, or deploys power is to say that in a significant way it shapes and constrains the field of possible actions of persons within some specific social context.
>
> J. Rouse, *Knowledge and power: toward a political philosophy of science*, 1987, pp. 210–11

Bruno Latour, an anthropologist of science, has considered the question of power in a way that is of particular relevance to our analysis of maps and theories. Power is not, as many believe, the cause of society. It is not the glue that bonds classes or groups together. Rather, according to Latour, it is the consequence of association, and it is the varying techniques of association that should be the focus of study, in looking at power. John Law, a sociologist of science who takes a similar approach to Latour's, has looked at the methods of long-distance control that were necessary for the Portuguese to sustain a trading route to India. Law concludes that the power of the Portuguese trading empire derived from the forms of association embodied in three essential ingredients: documents, devices and drilled personnel (John Law, 'On the methods of long-distance control', 1986, p. 234ff).

Latour has discussed the difference between what he calls 'savage' and 'civilised' geography in the context of searching for an explanation of the difference between what are often referred to as 'scientific' and 'primitive' cultures. This, he argues, must not be looked for in terms of some 'great divide' based on the postulation of radically different intellects, cultures or societies. Instead we have to look for small mundane differences.

10.1

La Pérouse travels through the Pacific for Louis XVI with the explicit mission of bringing *back* a better map. One day, landing on what he calls Sakhalin he meets with the Chinese and tries to learn from them whether Sakhalin is an island or a peninsular. To his great surprise the Chinese understand geography quite well. An older man stands up and draws a map of his island on the sand with the scale and the details needed by La Pérouse. Another, who is younger, sees that the rising tide will soon erase the map and picks up one of La Pérouse's notebooks to draw the map again with a pencil . . .

What are the differences between the savage geography and the civilized one? There is no need to bring a prescientific mind into the picture, nor any distinction between the close and open predicaments . . . nor primary and secondary theories . . . nor divisions between implicit and explicit, or concrete and abstract geography. The Chinese are quite able to think in terms of a map but also to talk about navigation on an equal footing with La Pérouse. Strictly speaking, the ability to draw and to visualize does not really make a difference either, since they all draw maps more or less based on the same principle of projection, first on sand, then on paper. So perhaps there is no difference after all and, geographies being equal, relativism is right? This, however, cannot be, because La Pérouse does something that is going to create an enormous difference between the Chinese and the European. What is, for the former, a drawing of no importance that the tide may erase, is for the latter the *single object* of his mission. What should be brought into the picture is how the picture is brought back. The Chinese does not have to keep track, since he can generate many maps at will, being

born on this island and fated to die on it. La Pérouse is not going to stay for more than a night; he is not born here and will die far away. What is he doing, then? He is passing through all these places, in order to take something *back* to Versailles where many people expect his map to determine who was right and wrong about whether Sakhalin was an isand, who will own this and that part of the world, and along which routes the next ships should sail. Without this peculiar trajectory, La Pérouse's exclusive interest in traces and inscriptions will be impossible to understand – this is the first aspect; but without dozens of innovations in inscription, in projection, in writing, archiving and computing, his displacement through the Pacific would be totally wasted – and this is the second aspect, as crucial as the first. We have to hold the two together. Commercial interests, capitalist spirit, imperialism, thirst for knowledge, are empty terms as long as one does not take into account Mercator's projection, marine clocks and their markers, copper engraving of maps, rutters, the keeping of 'log books,' and the many printed editions of Cook's voyages that La Pérouse carries with him. This is where the deflating strategy I outlined above is so powerful. But, on the other hand, no innovation in the way longitude and latitudes are calculated, clocks are built, log books are compiled, copper plates are printed, would make any difference whatsoever if they did not help to muster, align, and win over new and unexpected allies, far away, in Versailles.

(Bruno Latour, 'Visualisation and cognition', 1986, pp. 5–6)

The answer, he claims, lies in the power of techniques of writing and imaging. They do not achieve this power in and of themselves but as a result of their capacity to muster allies on the spot – allies, that is, in the struggle over what is to count as fact. To illustrate his argument he tells the story of La Perouse's encounter with the Chinese on Sakhalin Island. (Read ITEM **10.1**.) This story has parallels with that of the Inuit and the Belcher Islands (ITEMS **4.7–4.9**). Clearly the Hudson Bay Company acquired greater power than the Inuit through the production of a more powerful map.

Thus we can now see that the real distinguishing characteristic of Western maps is that they are more powerful than aboriginal maps, because they enable forms of association that make possible the building of empires (see ITEMS **10.2** and **10.3**), disciplines like cartography and the concept of land ownership that can be subject to juridical processes (ITEM **10.4**). Western and non-Western societies alike are based on knowledge networks, the important difference being in the mobility of the network. The Western one can be mobilised to cover the whole earth, if not the universe, whereas aboriginal ones are usually dependent on interpersonal oral modes of transmission. One of the most effective devices that Western maps employ in creating power is the grid, as we have seen in Exhibit 4. But, as we have also seen, the grid does not provide power of itself.

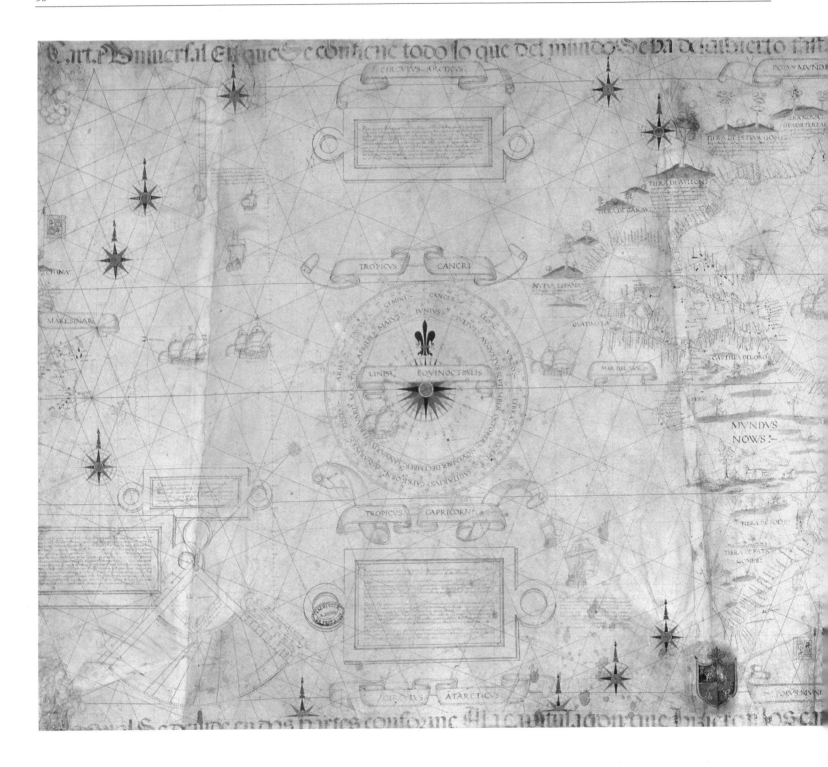

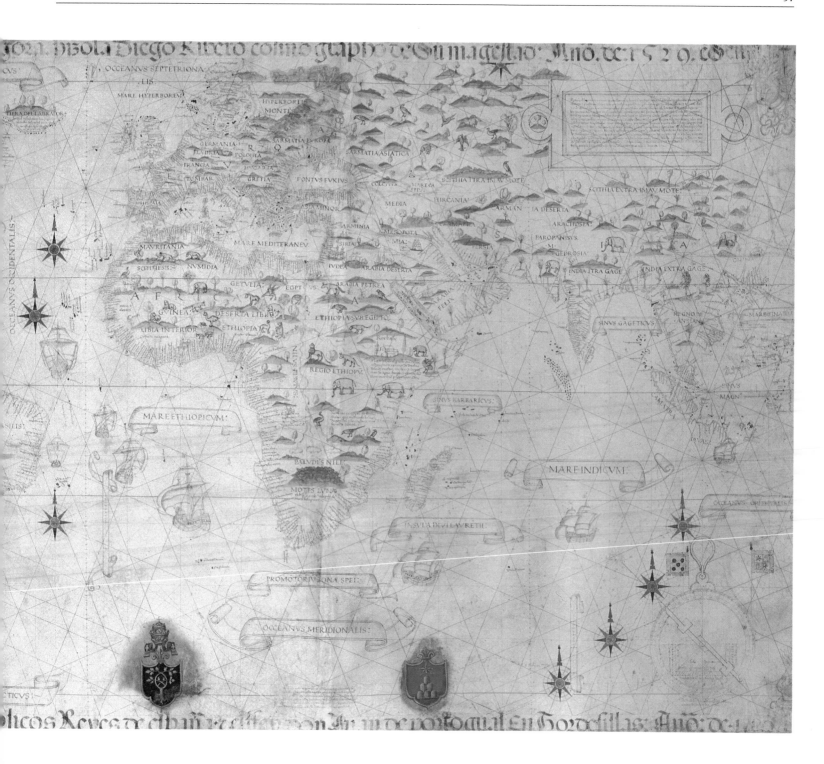

10.3
Augustus Caesar's edict on the 13th-century
Hereford *mappamundi* ordering a survey of the
whole world.

10.2 (previous page)
The Portugese cartographer Diego Ribero's map of
the world (1529), showing Pope Alexander VI's 'line
of demarcation', dividing the undiscovered world
between Spain and Portugal, following the Treaty of
Tordesillas in 1494. Despite the fact that no one
had either the instruments or the techniques to
locate or define the line with any accuracy, the mere
fact of having a map enabled a division of the world
with immense political ramifications. The division
was renegotiated in the Treaty of Saragossa in 1529.
Portugal paid Spain 350 000 ducats to move the line
in the Pacific further east in order to protect its
monopoly of the Spice Islands. This readjustment
gave the Portugese all of Australia bar a thin slice of
the east coast, though this was not recognised in the
treaty since Spain did not know of Australia's
existence and Portugal kept her knowledge secret.
The original demarcation line was established at the
longitude that separates Western Australia from
South Australia to this day. The Atlantic
demarcation line is indicated by the two small flags
at bottom centre and the Pacific demarcation flags
are below the Spice Islands in the open sea where
Australia was not known to exist at this time, at
least by the Spanish.

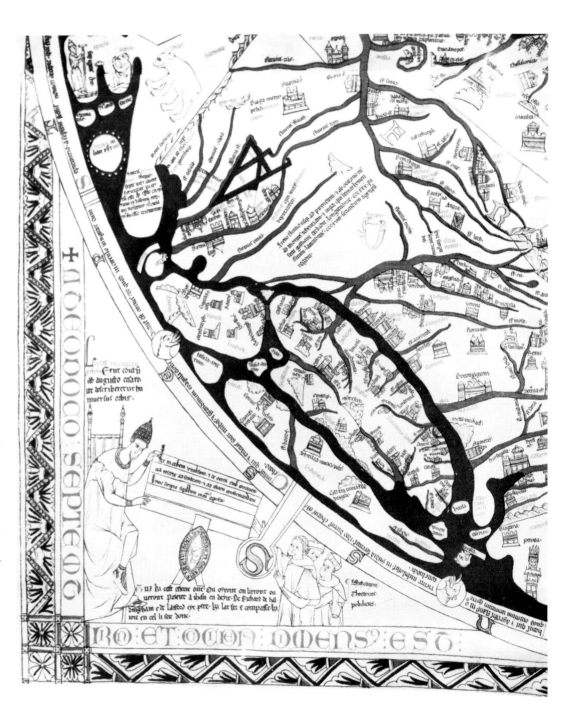

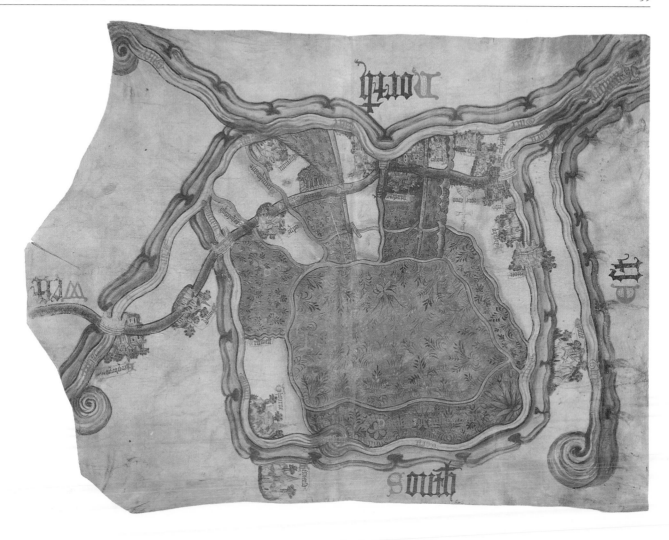

10.4

Map of Inclesmoor Yorkshire produced in a dispute during 1405–08 over rights to pasture and peat.

While Australian Aborigines are capable of mobilising their networks beyond the local circumstance, this mobility does have social and technical limitations. But interestingly, it is because of the common social element in networks of power that Aborigines are able to use maps of their country, as shown by the dreaming tracks, in supporting land claims in white courts of law (ITEM 10.5). Thus in certain circumstances maps can be mobilised to move from one network or form of life and inserted in another. It is a mobilisation not always guaranteed of success as shown by such aboriginal examples as Hawk Puma (ITEM 2.6), the Chippewa Indians (ITEM 4.1), Non Chi Ning Ga (ITEM 4.6) and the Mexican Indians (see back cover). Examination of this issue is beyond the limits of this book, but the way knowledge in maps relates to social and political power is an important area for further study.

10.5
Story of the Yalata map by Kingsley Palmer, Australian
Institute of Aboriginal Studies, Canberra.

The Southern Pitjantjatjara and the map of their Dreaming tracks

In 1981 as part of my research in the southern portion of the Great Victoria Desert I collected extensive information about the myths and tracks of the ancestral beings of the creative era of the Dreaming in this part of Australia. When my research was completed I drew up the Dreaming tracks and transferred this information onto a topographical map which provided a visual representation of the travels of the ancestral beings across this portion of Australia.

I subsequently returned to the community and took with me the map as a gift since I thought it was one way of demonstrating to the people something of the work that I had undertaken on their behalf. The response of the Aboriginal people was most interesting.

My Aboriginal helpers immediately regarded the map as an item of great cultural importance. They were concerned about the safety of the map and about the information that it contained. They considered that some of the details were in fact secret and should only be known by them. This was because a number of the myths included incidents that would generally only be discussed by mature and fully initiated men. They also recognized that the map contained secrets about their country and they were concerned to ensure that it did not fall into the wrong hands. They were particularly concerned about exploration geologists or certain government officers seeing the map and thereby understanding their secrets. It was agreed that the map would be placed in the vault of the bank in the nearby town. It was also arranged that the map could only be withdrawn with the permission of three of the older members of the community.

At this time the community was involved in extensive and protracted negotiation concerning the return of the Maralinga lands to the community. As a part of the negotiations a group of parliamentarians flew up from Adelaide to hold a meeting with the community at a site not far from the lands, in order to hear the Aboriginal claims to the country.

I was also invited to attend this meeting which was held in the sand dunes a few kilometres north of the trans-Australian railway line at a place known as Ooldea. When I arrived and the meeting commenced, I was surprised to discover that my map had been withdrawn from the bank. At a suitable time when the men had taken the parliamentarians to one side, the map was unrolled on the desert sands and the visitors were shown the Dreaming tracks as they extended over the countryside. The Aboriginal people were at great pains to point out the extent of the Dreaming tracks and the numerous sacred sites that were noted on the map and which were linked by the lines that represented the travels of the Dreaming ancestors.

Subsequently, in 1984, the majority of the Maralinga lands were handed back to the Aboriginal people who now hold this country freehold. The map has been returned to the bank where it has acquired the status of a title deed to the country. From the Aboriginal perspective it represents one proof of their close and enduring spiritual ties with their country. According to their understandings, the travels of the mythological beings and the sites they made are a spiritual reality which links them to the land and demonstrates their unity with it.

Exhibit 11 MAPS AND THEORIES CONCLUDED

'Only connect . . . '
(E. M. Forster, *Howards end*, epigraph)

What, then, have we learned about maps that is of some value in understanding theories? They are conventional, selective, indexical, embedded in forms of life, dependent on the understanding of a cognitive schema and practical mastery. They can be enormously powerful and can sustain not just successful exploration of foreign parts but whole empires. At base there is something more than merely metaphoric about maps and theories; they share a common characteristic which is the very condition for the possibility of knowledge or experience—connectivity. Since we cannot have a pure unmediated experience of our environment, that experience is better understood as an active construction resulting from a dialectical interaction between the 'lumps' in the landscape and our imposed connections of those lumps. Our experience and our representations are formative of each other and are only separable analytically. Hence there is an important sense in which the map *is* the territory, even though paradoxically the territory *is not* the map.

However, there are some difficulties in equating maps with theories if we take theories to be the embodiment of objective knowledge. This view of science has become problematic since the appearance of Thomas S. Kuhn's *Structure of scientific revolutions*. It is now recognised that theories and observations are inseparable, and also that for any given set of observations there exist in principle an indeterminately large number of theories that could fit. Most problematically though, theories do not come with a full set of rules about how to apply them in given cases. If you go back and reread the quote from Kuhn at the beginning, which was selected for its discussion of maps, you find Kuhn saying 'Through the theories they embody, paradigms prove to be constitutive of the research activity'. This gives the impression that to Kuhn theory is central to science. However, Kuhn himself in later trying to clarify his position makes it clear that he takes 'shared examples of practice' to be the central elements in science (T. S. Kuhn, 'Second thoughts on paradigms', 1977, pp. 459-99).

To see science as a 'field of practices' rather than a 'network of theories' makes a profound difference to our understanding. It is especially significant when it comes to maps. If maps are seen as theories in the sense of fully articulated objective knowledge, then only one small group of maps appears to qualify as real maps—the supposedly accurate contemporary Western maps. We have seen, in the process of looking at the exhibits, that there are difficulties with that position. On the one hand, it fails to acknowledge the workability and potential power of maps from non-Western cultures, while on the other hand, it fails to acknowledge the contingent character of Western maps. The approach we are considering here, by recognising maps as embodying shared examples of practice, makes it perfectly reasonable to accept all maps as having a local, contingent and indexical character intimately tied to human purposes and action.

The concept of science as fields of practice also highlights the importance of skills and tacit knowledge, which are often overlooked or suppressed when the purely theoretical is emphasised. Skills and tacit knowledge are modes of knowing the world that exemplify Wittgenstein's forms of life. They depend on givens that cannot be spoken of, in the same way that you cannot explain how to ride a bike. If we had to wait for a theoretical explanation of bike riding, nobody would even get on the saddle. If maps are shared examples of practice, perhaps science can be thought of as a compendia of maps, that is, an atlas, as an example of the way in which people have to work to make the whole hang together. Ultimately maps and theories gain their power and usefulness from making connections and enabling unanticipated connections. Science is an atlas not because all its theories are connected by logic, method and consistency. There is no such logic, or method or consistency. Science is riddled with contradiction and disciplinary division. Science is an atlas because the essence of maps and theories is connectivity. Maps and theories provide practical opportunities for making *connections* whenever and wherever it is socially and politically strategic.

FURTHER READING

L. Allen, *Time before morning, art and myth of the Australian Aborigines*, Crowell, New York, 1975

S. Alpers, 'The mapping impulse in Dutch art', in D. Woodward (ed.), *Art and cartography: six historical essays*, University of Chicago Press, Chicago, 1987

L. Bagrow, *History of cartography*, revised and enlarged R. A. Skelton, C. A. Watts, London, 1964

M. J. Blakemore, 'From way-finding to map-making: the spatial information fields of aboriginal peoples', *Progress in human geography*, vol. 5, 1981, pp. 1–24

M. J. Blakemore & J. B. Harley, 'Concepts in the history of cartography: a review and perspective', Monograph 26, *Cartographica*, vol. 17 no. 4, 1980

J. L. Borges, 'Of exactitude in science', in *A universal history of infamy*, tr. N. T. di Giovanni, Penguin Books, Harmondsworth, 1975

H. Brody, *Maps and dreams: Indians and the British Columbia frontier*, Penguin Books, Harmondsworth, 1983

G. Brotherson, *Image of the New World: the American continent portrayed in native texts*, Thames & Hudson, London, 1979

J. Burke, *The day the universe changed*, BBC, London, 1985

P. Carter, *The road to Botany Bay*, Faber & Faber, London, 1987

D. W. Chambers, *Beasts and other illusions* (Nature and human nature), Deakin University, Geelong, Vic., 1984

D. W. Chambers, *Imagining landscapes* (Nature and human nature), Deakin University, Geelong, Vic., 1984

D. W. Chambers, *Imagining nature* (Nature and human nature), Deakin University, Geelong, Vic., 1984

D. W. Chambers, *Is seeing believing?* (Nature and human nature), Deakin University, Geelong, Vic., 1984

D. W. Chambers, *Putting nature in order* (Nature and human nature), Deakin University, Geelong, Vic., 1984

D. Cosgrove & S. Daniels (eds), *The iconography of landscape*, Cambridge University Press, Cambridge, 1988

G. R. Crone, *Maps and their makers: an introduction to the history of cartography*, 5th edn, Dawson, Folkestone, Kent, and Archon, Hamden, Conn., 1978

C. Delano Smith, 'The emergence of "maps" in European rock art: a prehistoric preoccupation with place', *Imago mundi*, vol. 34, 1982, pp. 9–25

C. Delano Smith, 'Cartography in the prehistoric period in the old world: Europe, the Middle East, and North Africa', in J. B. Harley & D. Woodward (eds), *The history of cartography*, vol. 1 *Cartography in prehistoric, ancient, and medieval Europe and the Mediterranean*, University of Chicago Press, Chicago, 1987

S. Dewdney, *The sacred scrolls of the Southern Ojibway*, University of Toronto Press, Toronto, 1975

S. Edgerton, 'From mental matrix to *Mappamundi* to Christian Empire: the heritage of Ptolemaic cartography in the Renaissance', in D. Woodward (ed.), *Art and cartography: six historical essays*, University of Chicago Press, Chicago, 1987

L. Farrall, *Unwritten knowledge: case study of the navigators of Micronesia* (Knowledge and power), Deakin University, Geelong, Vic., 1984

C. O. Frake, 'Cognitive maps of time and tide among medieval seafarers', *Man*, vol. 20, 1985, pp. 154–70

F. Gale, 'Art as a cartographic form', *Globe: journal of the Australian Map Circle*, vol. 26, 1987, pp. 32–41

M. Gardner, *The annotated snark: the full text of Lewis Carroll's great nonsense epic 'The hunting of the snark'*, Penguin Books, Harmondsworth, 1967

A. Gell, 'How to read a map: remarks on the practical logic of navigation', *Man*, vol. 20, 1985, pp. 271–86

D. Gohm, *Antique maps of the Americas, West Indies, Australasia, Africa, the Orient*, Octopus Books, London, 1972

E. H. Gombrich, *Art and illusion: a study in the psychology of visual representation*, Pantheon, New York, 1960

H. Groger-Wurm, *Australian Aboriginal bark paintings and the mythological interpretation*, Australian Institute of Aboriginal Studies, Canberra, 1973

J. B. Harley, 'Maps, knowledge and power' in D. Cosgrove & S. Daniels (eds), *The iconography of landscape*, Cambridge University Press, Cambridge, 1988

J. B. Harley & D. Woodward (eds), *The history of cartography*, vol. 1 *Cartography in prehistoric, ancient, and medieval Europe and the Mediterranean*, University of Chicago Press, Chicago, 1987

P. D. A. Harvey, *The history of topographical maps: symbols, pictures and surveys*, Thames & Hudson, London, 1980

D. Hebdige, *Subculture: the meaning of style*, Methuen, London, 1979

A. G. Hodgkiss, *Understanding maps: a systematic history of their use and development*, Dawson, Folkstone, Kent, 1981

J. S. Hunter, 'The national system of scientific measurement', *Science*, vol. 210, 1980, pp. 869–74

J. S. Keates, *Understanding maps*, Longman, London, 1982

A. Korzybski, *Science and sanity*, 2nd edn, International Non-Aristotelian Library Publishing Co., Lancaster, Pa, 1941

T. S. Kuhn, *The structure of scientific revolutions*, 2nd edn, University of Chicago Press, Chicago, 1970

T. S. Kuhn, 'Second thoughts on paradigms', in F. Suppe (ed.), *The structure of scientific theories*, 2nd edn, University of Illinois Press, Champaign, 1977

B. Latour, 'Visualisation and cognition: thinking with eyes and hands', in H. Kuklick & E. Long (eds) *Knowledge and society: studies in the sociology of culture past and present*, vol. 6, JAI Press, Greenwich, Conn., 1986

B. Latour, 'The powers of association', in J. Law (ed.), *Power, action and belief: a new sociology of knowledge?*, Routledge & Kegan Paul, London, 1986

B. Latour, *Science in action: how to follow scientists and engineers through society*, Open University Press, Milton Keynes, 1987

J. Law, 'On the methods of long-distance control: vessels, navigation and the Portugese route to India', in J. Law (ed.), *Power, action and belief: a new sociology of knowledge?*, Routledge & Kegan Paul, London, 1986

J. Law, (ed.), *Power, action and belief: a new sociology of knowledge?*, Routledge & Kegan Paul, London, 1986

D. Lewis, 'Observations on route finding and spatial orientation among the Aboriginal peoples of the Western Desert region of Central Australia', *Oceania*, vol. 46, 1976, pp. 249–82

M. Lewis, 'The indigenous maps and mapping of North American Indians', *Map collector*, vol. 9, 1979, pp. 25–32

M. Lewis, 'Indian delimitations of primary biogeographic regions', in T. E. Ross & T. G. Moore (eds) *A cultural geography of North American Indians*, Westview Press, Boulder, Col., 1987

M. Lewis, 'The origins of cartography', in J. B. Harley & D. Woodward (eds), *The history of cartography*, vol. 1 *Cartography in prehistoric, ancient, and medieval Europe and the Mediterranean*, University of Chicago Press, Chicago, 1987

K. G. McIntyre, *The secret discovery of Australia: Portugese ventures 250 years before Captain Cook*, Pan Books, London, 1982

P. B. Medawar, 'Is the scientific paper a fraud?', *Listener*, vol. 70, no. 1798, September 1963, pp. 377–8

H. Morphy, '"Now you understand"–an analysis of the way Yolngu have used sacred knowledge to retain their autonomy', in N. Peterson & M. Langton (eds), *Aborigines, land and land rights*, Australian Institute of Aboriginal Studies, Canberra, 1983

F. Myers, *Pintupi country, Pintupi self*, Australian Institute of Aboriginal Studies, Canberra, and Smithsonian Institute Press, Washington, DC, 1983

R. Peters, 'Communication, cognitive mapping, and strategy in wolves and hominids', in R. L. Hall & H. S. Sharp (eds), *Wolf and man: evolution in parallel*, Academic Press, New York, 1978

N. Peterson, 'Totemism yesterday: sentiment and local organisation among the Australian Aborigines', *Man*, vol. 7, 1972, pp. 12–32

N. Peterson & M. Langton (eds), *Aborigines, land and land rights*, Australian Institute of Aboriginal Studies, Canberra, 1983

J. Piaget & B. Inhelder, *The child's conception of space*, W. W. Norton, New York, 1967

M. Polanyi, *Personal knowledge: towards a post-critical philosophy*, Routledge & Kegan Paul, London, 1958

K. Popper, *Unended quest: an intellectual autobiography*, Fontana, Glasgow, 1986

A. H. Robinson & B. B. Petchenik, *The nature of maps: essays toward understanding maps and mapping*, University of Chicago Press, Chicago, 1976

J. Rouse, *Knowledge and power: toward a political philosophy of science*, Cornell University Press, Ithaca, NY, 1987

M. Rudwick, 'The emergence of a visual language for geological science', *History of science*, vol. 14, 1976, pp. 149–95

C. Sagan, *The cosmic connection: an extra-terrestrial perspective*, Doubleday, New York, 1973

S. Shapin & S. Schaffer, *Leviathan and the air-pump: Hobbes, Boyle and the experimental life*, Princeton University Press, Princeton, NJ, 1985

R. W. Shirley, *The mapping of the world: early printed world maps, 1472–1700*, Holland Press, London, 1984

R. Temple, *The genius of China*, Simon & Schuster, New York, 1989

S. Toulmin, *The philosophy of science*, Hutchinson, London, 1953

H. Watson with the Yolngu community at Yirrkala & D. W. Chambers, *Singing the land, signing the land* (Nature and human nature), Deakin University, Geelong, Vic., 1989

E. Wells, *Reward and punishment in Arnhemland, 1962–63*, Australian Institute of Aboriginal Studies, Canberra, 1979

J. N. Wilford, *The mapmakers*, Knopf, New York, 1981

N. Williams, *The Yolngu and their land: a system of land tenure and the fight for its recognition*, Australian Institute of Aboriginal Studies, Canberra, 1986

L. Wittgenstein, *Philosophical investigations*, Blackwell, Oxford, 1958

D. Woodward, (ed.), *Art and cartography: six historical essays*, University of Chicago Press, Chicago, 1987

J. Ziman, *Reliable knowledge: an exploration of the grounds for belief in science*, Cambridge University Press, Cambridge, 1978

ACKNOWLEDGMENTS

Front cover, colour transparency courtesy of Österreichische Nationalbibliothek, Vienna; **title page**, L. A. Brown, *The story of maps*, Little, Brown & Co., Boston, 1950, p. 169; **imprint/contents pages**, map of Africa reproduced from Liverpool University Library copy, Ryl. N.2.24–6, by permission of the librarian; **opposite Preface**, map of Australia courtesy of National Library of Australia from J. J. Maslen, *The friend of Australia or, a plan for exploring the interior*, Hurst, Chana, London, 1830, Ferguson no. 1379; **1.1**, courtesy of Service Photographique, Bibliothèque Nationale, Paris; **1.2**, from *A universal history of infamy* by Jorge Luis Borges, tr. Norman Thomas di Giovanni (Penguin Books, 1973), © Emece Editores, S. A., and Norman Thomas di Giovanni, 1970, 1971, 1972, reproduced by permission of Penguin Books Ltd, London; **1.3**, from M. Gardner (ed.), *The annotated snark*, rev. edn, Penguin, London, 1974, by courtesy of Simon & Schuster, Inc. (proprietors); **2.1**(left), Mercator Map, 1569, from R. W. Shirley, *The mapping of the world*, Holland Press, London, 1984, courtesy of Service Photographique, Bibliothèque Nationale, Paris; **2.1**(right), outline of Mercator's world map from G. R. Crone, *Maps and their makers*, 5th edn, Dawson, Folkestone, Kent, 1978; **2.2**, **2.3**, Peters World Maps © Prof. Dr Arno Peters, used by permission of Oxford Cartographers, Eynsham, Oxford; **2.5**, MS Parm. 1614, reproduced by courtesy of Biblioteca Palatine di Parma (transparency: Vivi Papi); **2.6**, F. Guaman Poma de Ayala, *Nueva corónica y buen gobierno*, facsimile edn, Paris, 1936, reproduced from G. Brotherston, *Image of the new world*, Thames and Hudson, London, 1979, pp. 238–9; **2.7**, from the book *The cosmic connection*, by Carl Sagan, Anchor Press, Garden City, NY, © 1973 (photo: Jerome Agel); **2.8**, copyright London Regional Transport, Registered User no. 88/E/427; **3.1**, **3.2**, J. B. Harley & D. Woodward (eds), *The history of cartography*, vol. 1, *Cartography in prehistoric, ancient and medieval Europe and the Mediterranean*, University of Chicago Press, 1987, figs 4.28 and 4.29; **3.3**, from copy by Grace Huxtable in James Mellaart, 'Excavations at Catal Hüyük, 1963: Third preliminary report', *Anatolian Studies* vol. 14, 1964, pp. 39–119, pl.VI; **3.4**(left), P. D. A. Harvey, *The history of topographical maps*, Thames and Hudson, London, 1980, p. 125, fig.71; **3.4**(right), photo courtesy of Dr J. Oelsner, Friedrich-Schiller-Universität, Jena, DDR; **3.5**, photo

and line drawing from R. F. S. Starr, *Nuzi: Report on the excavations*, vol. 2, Harvard University Press, 1937, pl.55, items T, U, reproduced by permission; **3.6**, colour transparency and reproduction courtesy of Archivio di Stato di Venezia; **3.7**, from *Pages from the altar of Kiangsi province, south-east China, 18th century*, pl.V (MS 16356), courtesy of the British Library; **3.8**, colour transparency from Windsor Castle, Royal Library, © Her Majesty, Queen Elizabeth II; **3.9**, transparency from Cotton MS Augustus I ii 39, courtesy of the British Library; **4.2**, map drawn by Indians on birch-bark (shelf mark) BL RUSI (misc.)1, by permission of the British Library; **4.3**(top), Marshall Islands chart of coconut fibre and shells (British Museum no. 2289), photo © Trustees of the British Museum; **4.3**(bottom), Marshall Islands sailing chart of the *meddo* type, photo © Trustees of the British Museum; **4.4**(top), chart of peninsula between two fjords near Sermiligak, from Gustav Holm, Umiak Expedition to East Greenland 1884–85 (map La 21), reproduced by permission of the National Museum of Denmark, Department of Ethnography (photo: Lennart Larsen); **4.4**(middle and bottom), from Gustav Holm collection from East Greenland 1885 (maps La 19, 20), photos © the Greenland Museum 1988; **4.5**, from S. Dewdney, *The sacred scrolls of the Southern Ojibway*, University of Toronto Press, Toronto, 1975, reprinted by permission of the publisher; **4.6**, reproduced from *The Map Collector*, issue 9, 1979, fig. 3b, p. 27; **4.7**, Cartographic Branch, National Archives, Washington, DC, record group 75, map 821, tube 520; **4.8**, reproduced from *The Map Collector*, issue 9, 1979, fig. 4b, p. 27; **4.9**, photo courtesy of Butler Library, Columbia University in the City of New York; **4.10**, British Admiralty chart 863, Hudson Bay and Strait, 28 June 1884, large corrections May 1888, National Archives of Canada, C-106774; **4.11**, Admiralty chart no. 863, used by permission of the Canadian Hydrographic Office and the Hydrographer of the Navy, Ministry of Defence, Taunton, UK; **5.1**, reproduced by permission from N. Williams, *The Yolngu and their land*, Australian Institute of Aboriginal Studies, Canberra, 1986, endpapers (artist: M. McKenzie); **5.2**, bark painting, *Crocodile and fire dreaming*, Deakin University collection, reproduced by permission of Mr Galarrwuy Yunupingu, Yirrkala, NT, for the Gumatj clan; **5.4**, bark painting, *Dugong and fire dreaming*, Deakin University collection, reproduced by permission of Mr Galarrwuy Yunupingu, Yirrkala, NT, for the Gumatj clan; **5.5**, bark painting, *Water goannas and water dreaming*, Deakin University collection, reproduced by permission of Djamika; **6.1**, Ordnance Survey 1:63360, 7th series, sheet 173 (East Kent), edn A, 1959, crown

copyright reserved, reproduced by permission of Ordnance Survey (photo: The British Library); **6.2**, **6.3**, courtesy of Hunting Aerofilms; **7.1**, RGS 33898, 'World map of Ptolomy, Klaudius', transparency courtesy of Bridgman Art Library, London; **7.2**, J. B. Harley & D. Woodward (eds), *The history of cartography*, vol. 1, *Cartography in prehistoric, ancient and medieval Europe and the Mediterranean*, University of Chicago Press, 1987, fig.18.4, p. 297; **7.3**, Ebstorf world map, from L. Bagrow, *History of cartography*, rev. and enlarged by R. A. Skelton, C. A. Watts, London, 1964, pl.E; **7.4**, British Museum MS 30088, reproduced by permission of Trustees of the British Museum (colour transparency: British Library); **7.5**, D. Gohm, *Antique maps*, Octopus Books, London, 1972, p. 110; **7.6**, 'The Evil Genius of Europe' map of 1857, BL 1078 (24), by permission of the British Library; **7.7**, collection of the Worshipful Company of Makers of Playing Cards, Guildhall Library, City of London (photo: Godfrey New Photographics); **7.8**, 'The false map', 1854, from the Strachan & Co collection, University of Melbourne archives, by permission; **8.1**, E. Bullard, E. J. Everett & A. Gilbert Smith, *Philosophical Transactions of the Royal Society of London*, ser.A, vol. 258, 1965, by permission of the Royal Society; **8.2**, from Gilbert, *Geographical Journal*, vol. 124, 1958, p. 179, by permission of the Royal Geographical Society; **8.3**, from R. A. Phinney (ed.), *The history of the earth's crust*, fig.4, facing p. 80, copyright © 1986 by Princeton University Press, reprinted with permission of Princeton University Press; **9.1**, C. O. Frake, 'Cognitive maps of time and tide among medieval seafarers', *Man*, vol. 20, 1985, fig.2, by permission of the Royal Anthropological Institute of Great Britain and Ireland; **9.2**, D. Gohm, *Antique maps*, Octopus Books, London, 1972, p. 95; **9.3**, by permission of AGPS, Canberra; **10.1**, B. Latour, *Knowledge and society: Studies in the sociology of culture past and present*, vol. 6, JAI Press, Greenwich, CT, by permission of the publisher; **10.2**, Diego Ribero's map of the world, 1529, courtesy of Bibliotheca Apostolica Vaticana; **10.3**, by permission of the Dean and Chapter of Hereford Cathedral; **10.4**, Crown Copyright, Public Record Office MPC 56 ex DL 31/61; **10.5**, Kingsley Palmer, Australian Institute of Aboriginal Studies, Canberra.